T0235664

Indigenous Peoples, Natural Resources and Governance

This book offers multidisciplinary perspectives on the changing relationships between states, indigenous peoples and industries in the Arctic and beyond. It offers insights from Nordic countries, Canada, Australia, New Zealand and Russia to present different systems of resource governance and practices of managing industry-indigenous peoples' relations in the mining industry, renewable resource development and aquaculture.

Chapters cover growing international interest on Arctic natural resources, globalization of extractive industries and increasing land use conflicts. It considers issues such as equity, use of knowledge, development of company practices, conflict-solving measures and the role of indigenous institutions.

- Focus on Indigenous peoples and Governance triangle
- Multidisciplinary: political science, legal studies, sociology, administrative studies, Indigenous studies
- Global approach: Nordic countries, Canada, Russia, Australia, New Zealand and Canada
- Thorough case studies, rich material and analysis

The book will be of great interest to legal scholars, political scientists, experts in administrative sciences, authorities at different levels (local, regional and nations), experts in human rights and natural resources governance, experts in corporate social governance.

Monica Tennberg, Research Professor, Arctic Centre, University of Lapland, Finland, Professor II, Centre for Sami Studies, UiT The Arctic University of Norway.

Else Grete Broderstad, Professor, Centre for Sami Studies, UiT The Arctic University of Norway.

Hans-Kristian Hernes, Professor, Department of Social Sciences, UiT The Arctic University of Norway.

Routledge Research in Polar Regions

Series Editor: Timothy Heleniak,
Nordregio International Research Centre, Sweden

The Routledge series in Polar Regions seeks to include research and policy debates about trends and events taking place in two important world regions: the Arctic and Antarctic. Previously neglected periphery regions, with climate change, resource development and shifting geopolitics, these regions are becoming increasingly crucial to happenings outside these regions. At the same time, the economies, societies and natural environments of the Arctic are undergoing rapid change. This series seeks to draw upon fieldwork, satellite observations, archival studies and other research methods which inform about crucial developments in the Polar regions. It is interdisciplinary, drawing on the work from the social sciences and humanities, bringing together cutting-edge research in the Polar regions with the policy implications.

For more information about this series, please visit: www.routledge.com/ Routledge-Research-in-Polar-Regions/book-series/RRPS

Indigenous Peoples, Natural Resources and Governance

Agencies and Interactions

Edited by Monica Tennberg,
Else Grete Broderstad and
Hans-Kristian Hernes

LONDON AND NEW YORK

First published 2022
by Routledge
2 Park Square, Milton Park, Abingdon, Oxon OX14 4RN

and by Routledge
605 Third Avenue, New York, NY 10158

Routledge is an imprint of the Taylor & Francis Group, an informa business

© 2022 selection and editorial matter, Monica Tennberg, Else Grete Broderstad and Hans-Kristian Hernes; individual chapters, the contributors

British Library Cataloguing-in-Publication Data
A catalogue record for this book is available from the British Library

Library of Congress Cataloging-in-Publication Data
A catalog record has been requested for this book

ISBN: 978-0-367-67415-1 (hbk)
ISBN: 978-0-367-67416-8 (pbk)
ISBN: 978-1-003-13127-4 (ebk)

DOI: 10.4324/9781003131274

Typeset in Bembo
by SPi Technologies India Pvt Ltd (Straive)

Contents

Illustrations

Figures

Tables

Contributors

Emma Borg is an Assistant Researcher at the University of Helsinki and a recent Master of Laws graduate. Her experience and interests lie within international law, human rights and Arctic policy.

Camilla Brattland is an Associate Professor at the Department of Social Sciences, UiT – The Arctic University of Norway. Her research focuses on coastal Sámi use of marine resources, traditional ecological knowledge and Sámi participation in natural resource management, where the mapping of traditional use is among her methods. As part of the Arctic Governance Triangle project (TriArc), she analyzes interactions between Indigenous peoples, the aquaculture industry and the state, with a comparative view of Indigenous contexts in Norway, Canada and New Zealand.

Else Grete Broderstad holds a doctorate in Political Science, is a Professor in Indigenous Studies and coordinates the Indigenous Master's Programme at the Centre for Sami Studies, UiT – The Arctic University of Norway. Her research areas include Indigenous rights and political participation as well as governance differences and similarities in the Circumpolar North. She is currently leading research projects on resource management and conflicting interests between Indigenous traditional livelihoods and large-scale industries.

Dorothée Cambou is an Assistant Professor in Sustainability Science at the University of Helsinki (Faculty of Law and the Helsinki Institute of Sustainability Science, HELSUS). The main focus of her current research lies in international law and human rights and, more particularly, on the rights of Indigenous peoples, including those of the Sámi. Her expertise is in the field of land and resource governance, self-determination issues, business and human rights as well as Arctic studies.

Kaja Nan Gjelde–Bennett received her MA in Indigenous Studies (2020) from UiT – The Arctic University of Norway as a U.S.–Norway Fulbright Grantee and a Norway–America Association Scholar. She completed her BA in Global Studies and Hispanic Studies (2017) at Pacific Lutheran University. As an independent writer and researcher, Gjelde-Bennett has participated in several scholarly projects through the Centre for Women's and Gender Research and the Centre for Sámi Studies at UiT, including the TriArc Project.

Currently pursuing a Ph.D. in Spanish and Portuguese at the University of New Mexico, Gjelde-Bennett's research interests include Indigenous methodologies, Indigenous rights, language revitalization, intersectional feminism and contemporary Hispanic literature.

Marina Peeters Goloviznina is a doctoral fellow at the Centre for Sami Studies, UiT – The Arctic University of Norway. She gained her MA in Sociology (2001) and a Licentiate in Sociology (kandidat nauk) (2005) from the European University at Saint-Petersburg (Russia). Her research interests lie in critical Indigenous studies, social norms research and the politics of recognition. In particular, she studies issues relating to Indigenous institutions, agency and rights.

Hans-Kristian Hernes is a Professor in Political Science at UiT – The Arctic University of Norway, and currently deputy chair in the Department of Social Sciences. He teaches courses in public policy and political theory, and is involved in courses focusing on the role of Indigenous peoples. His research explores the role of interest groups, fisheries politics and resource management, and he is currently involved in research on large industrial projects in Indigenous areas and the transformation to new renewable energy in remote areas.

Catherine Howlett is currently the Director of Higher Degree Research Training at Gnibi College of Indigenous Knowledge at Southern Cross University in Lismore, Australia. She is also affiliated with the Centre for Sami Studies, UiT – The Arctic University of Norway through two research projects. With a doctorate in Politics and Public Policy, she is involved in several international research projects on Indigenous peoples and resource development and management. She is currently working with colleagues at Wollongong University on developing a program to include Indigenous Knowledge into land and marine decision-making processes.

Rebecca Lawrence is an Associate Professor at Sydney Environment Institute, University of Sydney, and in the Department of Political Science, Stockholm University. She has worked with Indigenous Sámi communities in Sweden to protect reindeer grazing lands from wind power, forestry and mining projects, and has been affiliated with the Centre for Sami Studies, UiT – The Arctic University of Norway through the IndKnow project.

Horatio Sam-Aggrey holds a doctorate in sociology and a Master's degree in Indigenous and Northern Studies. He has conducted research on northern and Indigenous issues for over six years and previously worked on Aboriginal engagement to facilitate the deployment of various power facilities in Canada's North. Most recently, he worked as a Research Assistant at the Centre for Sami Studies at UiT – The Arctic University of Norway. He was additionally a Professional Research Associate with the University of Saskatchewan, Canada. Sam-Aggrey is currently a Project Assessment Analyst in northern Canada. His research interests include traditional knowledge, Indigenous consultations, climate change and environmental assessment, and nuclear/radiation perception studies.

Per Sandström is an Associate Professor at the Swedish University of Agricultural Sciences in the Department of Forest Resource Management, Umeå. He leads and participates in a number of projects assessing multiple land use impacts and planning, typically in relation to reindeer husbandry. His research covers diverse and often multidisciplinary approaches incorporating both scientific and Indigenous and local knowledge systems aimed at resolving land use conflicts. He is affiliated with the Centre for Sami Studies, UiT – The Arctic University of Norway through the TriArc project.

Anna Skarin is a Professor in Reindeer husbandry at the Swedish University of Agricultural Sciences, with a doctorate in Animal Science. Her research seeks to understand drivers of animal behavior and habitat selection and specifically how humans and infrastructure impact reindeer behavior and use of the landscape. She has long worked on reindeer movement and habitat selection, and hers was one of the first studies with GPS-collared reindeer. Skarin has also pioneered reindeer habitat selection analysis in relation to wind farms.

Gabrielle A. Slowey is the former Director of the Robarts Centre for Canadian Studies and an Associate Professor in the Department of Politics at York University, Toronto, where she teaches courses in Canadian, Indigenous and Arctic Politics. With a doctorate in Political Science, Slowey investigates the intersection between governance, resource extraction, economic development, the environment and the state in multiple regions (Northern Alberta, Northern Quebec, Yukon, Northwest Territories, Ontario, the United States, Australia and New Zealand). She is the author of *Navigating neoliberalism: self-determination and the Mikisew Cree First Nation* (UBC Press, 2008).

Monica Tennberg holds a doctorate in political sciences and leads the Northern Political Economy research group at the Arctic Centre, University of Lapland, Finland, as a research professor. She also served as professor II (20%) at the Centre of Sami Studies, UiT The Arctic University of Norway, during 2016–2018. Her current research focuses on Foucauldian critical perspectives to Arctic politics, in particular urban development, water governance, climate change and sustainable development.

Acknowledgments

We would like to thank the Norwegian Ministry of Foreign Affairs, Section for the High North, Polar Affairs and Resources for funding the Focal Point North project (grant no. RER-12/0073), which made this book project possible. Thanks are also due to the Research Council of Norway for financing research on The Arctic governance triangle: government, Indigenous peoples and industry in change (research grant no. 259416). This project has given rise to many of the chapters in this book. Also, we are grateful to the Centre for Sami Studies at UiT – The Arctic University of Norway for administrative and financial support. The publication charges for this book have been funded by a grant from the publication fund of UiT The Arctic University of Norway.

We particularly acknowledge the skilled assistance of Pirkko Hautamäki in proofreading our work. We are grateful to Camilla Brattland and Arto Vitikka for their expertise in the art of map-making, and appreciate the work of the anonymous reviewers who have offered valuable feedback and comments on the chapters.

The author of Chapter 2 is grateful to her TriArc colleagues for constructive discussions, and to Eva Maria Fjellheim and Torvald Falch for good comments on the chapter draft. The authors of Chapter 3 thank Inger-Ann Omma from Vapsten sameby for valuable input into the research. Anna Skarin and Per Sandström are grateful for the support of the Swedish Energy Agency and the research grant no. 46780-1. The authors of Chapter 4 would like to thank the following for participation in the research: Silje Karine Muotka, Elina Hakala and Lise Bergan and inhabitants from the Strándda community in Norway; representatives of the various Iwi who participated in interviews for this research in New Zealand, representatives of New Zealand King Salmon and Justine Inns from Ocean Law NZ, for taking time out to detail the nuances of the NZ case. The author of Chapter 5 expresses her greatest debt to her informants from the Republic of Sakha (Yakutia), who generously shared their thoughts, experiences and time and warmly welcomed her in their homes and workplaces. She expresses her sincere gratitude to Maria Petrovna Pogodaeva, who taught her much about *obshchiny* in the Sakha Republic. The author of Chapter 7 wishes to acknowledge Graham White for his editorial review and comments on a earlier version of the chapter, the editors for inviting her to participate in this volume and of course the First Nation communities in northern Ontario and the leaders with whom she has had the pleasure of working with (including Alvin Fiddler and the James Bay Treaty Conference).

Finally, as editors we are grateful to the authors and their respective research participants for sharing valuable knowledge making this book possible.

Abbreviations

AIPON	Association of Indigenous Peoples of the North of the Republic of Sakha (Yakutia)
ASC	Aquaculture Stewardship Council
AZRF	Arctic Zone of the Russian Federation
CAB	County Administrative Board (Sweden)
CAD	Canadian dollar
CANZUS	Canada, Australia, New Zealand and the USA
CEAA	Canadian Environmental Assessment Agency
CEO	Chief executive officer
CERD	United Nations Committee on the Elimination of Racial Discrimination
CINAC	Crown–Indigenous relations and Northern Affairs Canada
CLCA	Comprehensive land claims agreement (Canada)
CSR	Corporate social responsibility
EA	Environmental assessment
EE	Etnologicheskaya ekspertiza (comparable to a Social Impact Assessment) (Russia)
EIA	Environmental impact assessment
EITI	Extractive Industries Transparency Initiatives
EMAB	Environmental Monitoring Advisory Board (Canada)
EU	European Union
FOR	Fred Olsen Renewables
FPIC	Free, prior and informed consent
FPN	Focal Point North
FSC	Forest Stewardship Council
GDP	Gross domestic product
GNWT	Government of Northwest Territories (Canada)
GTA	Greater Toronto Area
HRC	Human Rights Committee
IBA	Impact benefit agreement
ICCPR	International Covenant on Civil and Political Rights
IFC	International Finance Corporation performance standards
IGO	Intergovernmental organization

ILO	International Labour Organization
ILUA	Indigenous Land Use Agreement (Australia)
IPO	Indigenous peoples' organizations
IR	International relations
ISC	Indigenous Services Canada
KMNS	Korrennye malochislennye narody Severa, Sibiri i Dal'nego Vostoka (Indigenous small-numbered peoples of the North, Siberia and the Far East) (Russia)
MPI	Minister for Primary Industries
MVEIRB	Mackenzie Valley Environmental Impact Review Board (Canada)
MVRMA	Mackenzie Valley Resource Management Act (Canada)
MZO	Minister's Zoning Order (Ontario, Canada)
NAN	Nishnawbe Aski Nation (Canada)
NCA	Norm contestation analysis
NGO	Nongovernmental organization
NNTT	National Native Title Tribunal (Australia)
NPO	Non-profit organization (Russia)
NRL	Norske Reindriftssamers Landsforbund (Sámi Reindeer Herders Association of Norway/Norgga boazosápmelaččaid riikkasearvvi)
NSC	Nordic Sámi Convention
NTA	Native Title Act 1993 (Australia)
NVE	Norwegian Water Resources and Energy Directorate
NWT	Northwest Territories (Canada)
NZ	New Zealand
NZKS	The New Zealand King Salmon Co. Ltd
OED	Ministry of Petroleum and Energy (Norway)
OIPR	Ombudsman for Indigenous peoples' rights, the Republic of Sakha (Yakutia)
OVOS	Otsenka vozdeystviya na okruzhaiyschuiy sredu (comparable to an Environmental Impact Assessment) (Russia)
RCN	Research Council of Norway
RMA	Resource Management Act, New Zealand
RMG	Raw Materials Group
RS (Ya)	Republic of Sakha (Yakutia)
RTN	Right to negotiate (Australia)
SIA	Social impact assessment
SLO	Social license to operate
SRC	Sámi Rights Committee (Norway)
SSR	Svenska Samernas Riksförbund (National Union of the Swedish Sami People/Sámiid Riikkasearvi)
TOKM	Te Ohu Kaimoana (New Zealand)
TriArc	The Arctic Governance Triangle: government, Indigenous peoples and industry in change
TTNU	Territory of traditional nature use (Russia)
UNCED	United Nations Conference on Environment and Development

UNDRIP	UN Declaration on the Rights of Indigenous Peoples
UNO	Union of the Nomadic Obshchiny (Russia)
W&J	Wangan and Jagalingou peoples (Australia)
WLWB	Wek'èezhìi Land and Water Board
WRH	World Reindeer Herders Association

1 Indigenous rights and governance theory

An introduction

Hans-Kristian Hernes, Else Grete Broderstad and Monica Tennberg

Introduction

Indigenous peoples worldwide experience great tensions with extractive industries over resources and territories. Such tensions over large industrial projects are not new. Modern history is filled with stories of intrusion, dispossessed lands and destroyed possibilities for pursuing traditional economies and cultures. The current argument is that the pressure is increasing, conflicts are becoming more intense and extending to new and promising areas (such as the Arctic) and including new industries (such as renewable energy and aquaculture).

If not entirely disputed, this view is at least modified by those arguing that Indigenous peoples have got better rights and have become more equal partners through participation and sharing resources. The new instruments developed internationally by market actors or government bodies make it possible to deal with the often stalemate relationship between Indigenous groups and industries (Owen and Kemp, 2017). The gradual recognition of Indigenous peoples' rights includes participation in decision-making by states, direct negotiations with companies and possible economic benefits for Indigenous groups (O'Faircheallaigh, 2013, 2016).

Indigenous groups used to manage the pressure on land and other resources by appealing to state authorities with the expectation that the government would have resources to challenge industrial projects and companies and adopt necessary legal regulations to protect traditional Indigenous livelihoods. As the main actors in the international arenas, governments are also responsible for implementing international law in domestic settings. However, states—like big companies—have a dubious reputation among Indigenous peoples and are not always seen as the best protector of their rights and heritage.

Indigenous peoples take different roles in the life course of industrial projects. In their cooperation with the state and big companies, they are likely to face conflict and heated discussions over resources and the right to participate. What are the roles, then, that Indigenous peoples can assume, and are they co-opted victims rather than real participants? New regulations, whether created by the market or international law, leave room for Indigenous agency, but what kind of agency is it? As large projects will remain on the agenda and conflicts are bound to emerge, how can Indigenous peoples deal with the situation?

DOI: 10.4324/9781003131274-1

Awareness of these issues was on the rise at the Centre for Sami Studies at UiT The Arctic University of Norway some ten years ago. The attention was formalized in the Focal Point North project funded by the Ministry of Foreign Affairs. The project introduced students to increasing conflicts over natural resources in the Circumpolar North. It also enabled networking among researchers and made it clear that resource extraction was a main driver impacting Indigenous rights to land and resources. Several adjunct professor positions were affiliated to FPN, including research professor Monica Tennberg, one of the editors of this book. An outcome of these networks and discussions was the project *Arctic governance triangle: government, Indigenous peoples and industry in change* (TriArc) which was funded by the Research Council of Norway.

The goal behind the TriArc project was to examine challenges between large industrial development projects and traditional uses of land and other natural resources, and to study the governance arrangements which were to regulate the relationship. Among the starting points was an observation of conflicts and challenges of legitimacy, but also cases where industries and Indigenous peoples had managed to find platforms for reciprocal cooperation. A question was how the development of new regulations and mechanisms worked, and if Indigenous peoples were included in the processes. The project members also wanted to analyze the ways in which Indigenous involvement in processes of natural resource development was guided by international and national political and legal realities, the behavior of various corporate actors and Indigenous peoples' own institutions. To what extent could we identify forms of governance that promoted Indigenous engagement with natural resource development and management?

The intent then was to study institutional solutions at the local level, to clarify whether decisions were decentralized and had an element of inclusion and participation, and if—and how—frameworks at different levels (national or international) mapped out the development of the different institutions. In addition to studying the linkage between different levels, we aimed at a comparison between countries to grasp how different settings affected projects involving industry and Indigenous peoples.

The theoretical framework came from governance theory and the idea that governance processes involving actors in government (state), market and civil society could be illustrated in a (governance) triangle. We recognized that Indigenous peoples' governance was undergoing major changes: many premises were emerging from international processes and arenas, governments were increasingly including Indigenous institutions and organizations in decision-making, and there might be a move from governance by state (hierarchy and coordination) to other types of governance by market and civil society. The project defined civil society as local communities in general and Indigenous peoples as rights holders in particular. The use of several terms for market actors—business, company, business organizations, industries—reflects the variety of actors and also the multidisciplinary approaches in the project.

The rest of this chapter is organized into three main parts. The first, on Indigenous governance, covers some of the main elements in the development of

Indigenous rights during the last decades. The second part discusses Professor of Public Organization and Management Jan Kooiman's governance theory, and the third section introduces the different case studies presented in this book.

Indigenous governance

A turn from definitive rights

According to legal scholar James Anaya (2004), Indigenous peoples' rights are part of the development of human rights after World War II, with a shift from individuals to rights for groups. While former colonies became new independent nation states, it was the framework of established nation states in which most Indigenous peoples had to secure their rights as peoples. Early attempts at recognition of Indigenous rights were characterized by one-way processes in the sense that rights were "given" from the top, by state authorities. Another element was that the rights were considered as final and represented a definitive solution settling the relationship between the majority and minority groups.

Political theorist James Tully (2004), however, postulates that there has since been a change, a turnaround where rights develop in stages—and that they are fluid and changing in the midst of societal processes. When a group receives recognition, others will mobilize to oppose this or to achieve rights themselves. This can lead to a decline, but also to a gradual and continuous development and extension of rights. Furthermore, the processes are characterized by interaction: rights are not granted from above, but are developed in various forms of dialogue between actors so that those who fight for recognition are also involved (Tully, 2004). Such an understanding implies that other types of processes are required to ensure legitimacy, that the legitimacy of rights can be challenged, and that rights and institutions will undergo changes so that, for example, the content of self-determination will change.

Tully's point can be perceived to apply within a nation state through, for instance, political decision-making and court decisions. At the same time, increased activity in international arenas and the development of rights by international organizations is also a dynamic feature. In the United Nations, Indigenous peoples' rights are interpreted and reinterpreted by committees, which have created new premises in the domestic discussions of rights.

Turning to multilevel governance

For decades, Indigenous peoples from different parts of the world have worked to develop alliances, with researchers as key players, to gain recognition. Central issues were related to self-determination, protection of culture and to securing the basis for traditional industries. The most prominent of these processes led to the UN Declaration on the Rights of Indigenous Peoples (UNDRIP), adopted in 2007. While the declaration is not binding on individual states, it is nevertheless important given the strong support by the UN and is valued as an important symbol of the recognition of Indigenous peoples' position.

The UN Declaration on the Rights of Indigenous Peoples (UNDRIP) was in 2007 seen as a landmark in the work to strengthen the role of Indigenous people's vis-à-vis the government and to define important means for self-determination. The Indigenous and Tribal Peoples Convention (ILO Convention No. 169) and the International Covenant on Civil and Political Rights (ICCPR), in particular Article 27, also have a significant bearing on the premises for Indigenous peoples' rights. These conventions and declarations illustrate the efforts made by Indigenous peoples to "seek justice in international law" (Barelli, 2016).

International law can be loosely linked to nation states and the policies they choose to pursue. A distinctive feature of the Indigenous sphere is a clearer institutionalization of governance that binds different institutional levels and institutions together. One is a political dimension, with an emphasis on participation and involvement. The UN is a central arena where Indigenous peoples can meet: not only are they members of nations' delegations, but they also meet as independent (Indigenous) peoples, as is the case in the UN Permanent Forum on Indigenous Issues (Dahl, 2012). A parallel development has led to the establishment of other forums that strengthen the legal aspects through monitoring and development of guidelines for international conventions and declarations. This gives Indigenous peoples a stronger position than if the implementation were left to nation states alone.

Clarifications and interpretations are not without significance. It is through international work that Indigenous peoples—and nation states—have agreed on key mechanisms for their involvement. Based on the premise that Indigenous peoples are equal "peoples," the point of consultations and schemes such as "free, prior and informed consent" (FPIC) is to ensure that Indigenous peoples have the opportunity to exert real influence. Consultations signify a breach of traditional hierarchical management and entail that the authorities give Indigenous peoples a genuine opportunity to participate in decisions that affect them. Also, consultations "shall be undertaken, in good faith and in a form appropriate to the circumstances, with the objective of achieving agreement or consent to the proposed measures" (ILO C169, Article 6.2). Consultations take place between two peoples: Indigenous peoples and the majority peoples represented by the state.

Consultations are an important tool in the UN Declaration on the Rights of Indigenous Peoples too, and although the declaration is non-binding, the geographical scope is larger than ILO-C 169 (1989), which has been ratified by relatively few countries. The description of consultations primarily points to the responsibility of states to facilitate and implement, and the implications are not necessarily easy to detect. The principle of FPIC is more visible, more easily understood, and has to a greater extent than consultations emerged as a visible signal of the necessary premises for the involvement of Indigenous peoples in decision-making. So, in addition to governmental processes, FPIC has gained access to business organizations and, for example, environmental groups.

Implementation gap and local variations

The clear focus on international processes and arenas suggests standardization and equality between Indigenous peoples in different parts of the world, but the actual situation is different. While it is true that several states have implemented consultation schemes, Indigenous peoples' opportunities to participate and influence differ a great deal (Pirsoul, 2019). The UN Declaration on the Rights of Indigenous Peoples is admittedly highlighted as a central premise and requires domestic implementation, but real changes are easily counted, and efforts for implementation have been met with critique and opposition. Moreover, even if Indigenous peoples' rights are linked to developments in human rights—themselves widely supported—there is a considerable gap between any awareness and real support. The status of Indigenous peoples' rights in Sweden, for example, has been described as "organized hypocrisy" (Mörkenstam, 2019), nor have the Nordic countries been able to agree on a joint Sámi convention.

An important point in all of this is that the implementation of Indigenous peoples' rights that does take place varies significantly, and a range of actors have assumed leading roles in such implementation. Such variation stems from the different institutional features of the nation states, where there may be clear differences between unitary states such as Norway, Sweden and Finland and federal states such as Australia and Canada. In federal states, courts have played an important role in promoting implementation of Indigenous peoples' rights, while political processes have so far been the central path in the Nordic countries. Perhaps this is about to change through new court processes and decisions, as recent rulings in Sweden have demonstrated. At the same time, there are also differences in the legal and institutional position of Indigenous peoples. In contrast to the Nordic countries, for example, Canadian Indigenous peoples have had better control over territories through agreements with the authorities and security from the courts. In combination with the federal structure, this has facilitated land claim agreements unlike unitary states without local resource control.

Business and human rights

The business community is increasingly being challenged to respect human rights, and this is important in the context of Indigenous peoples too. The use of FPIC in business guidelines is an example (Wilson, 2016), but similarly relevant are corporate social responsibility (CSR) and social license to operate (SLO). Corporate social responsibility refers to companies' own ethical guidelines and principles to which adherence is expected, while SLO has a dynamic element in that businesses establish a relationship with local communities in order to gain acceptance for their operations. The degree of acceptance can vary, and there are also cases of overwhelming local support where the companies and local communities have overlapping interests. As a concept, however, SLO is not clearly defined, and its use probably depends on the geographical context. For example, it is so far rather irrelevant in northern Europe (Koivurova et al., 2015). Impact benefit agreements (IBAs) are—in some settings—used as the main tool to mitigate impacts and divide benefits from project development. They may be part of an SLO process and can be an effective way to provide payment to local communities. There is, however, a comprehensive debate over challenges related to objectives, social justice issues, state-Indigenous relationships and best practices for IBAs (e.g. Bradshaw et al., 2019; Cascadden et al., 2021), and to tie benefit sharing to parts of international law (biodiversity, human rights) (Morgera, 2016).

Governance

The development of governance theory as related to Indigenous peoples' rights to resources—which was the basis of the TriArc project—is part of a comprehensive change in perspectives on societal governance. Since the 1980s, there have been major changes in corporate governance and in perceptions of what constitutes good governance (Bevir, 2012). The postulate, or slogan, of "governance without government" illustrates a turn in which governance is no longer perceived as the domain of the state and where hierarchy is supplemented with other facets of governing. This does not necessarily mean that the state is completely absent. In many process and decisions, governments will remain a strong player, albeit with a different role, and other players in the market and civil society have become more prominent, setting the agenda and developing institutional solutions (Kooiman, 2003, p. 3).

The Forest Stewardship Council (FSC) labeling scheme, for example, was created in a collaboration between environmental organizations and industry because states had failed to agree on schemes to ensure sustainable forest management. Such "private" solutions are nevertheless not the dominant element in today's governance. We may be able to identify entirely public solutions but these do not necessarily follow formal lines. The concept of multilevel governance had an important foundation in studies of developments in the EU with interaction across different governmental levels in the public sector (Piattoni, 2009). Political scientist James Rosenau (1997) discusses the ways in which globalization has challenged the boundaries between local, national and international politics and created new meeting places outside established formal arenas. Such changes have challenged the nation state's dominant role in governance, but the authorities are still among the key players.

Corporate social responsibility and social license to operate schemes thus illustrate attempts to establish management on the basis of a direct relationship between companies and civil society actors. What still remain as a state responsibility are consultations, which differ from previous management praxis in that hierarchy is to be replaced by interaction grounded in an equal partnership between Indigenous peoples and the authorities.

We can approach governance (without government) in different ways, as is illustrated by a rich body of research literature. The development reflects a need in society to govern in new ways and make room for increased flexibility, involvement of various actors and fewer elements of hierarchy. Such governance has, for example, been argued to be more efficient and increase legitimacy to a greater extent than traditional government-defined governance (Dryzek, 1999; Young, 1999). Various governance schemes have also created fertile ground to develop arenas for co-production and co-creation as measures for innovation and change in the public sector (Torfing et al., 2019).

The approach in this book is guided by an understanding of governance as developed by Kooiman (2003), which has been helpful in studies of marine resource management (Kooiman and Bavinck, 2013; Jentoft and Chuenpagdee, 2015; Kooiman et al., 2005). The approach is especially useful compared to other research on interactive governance where the purpose is primarily on the study of changes in administrative and political structures (Torfing et al., 2019). The work by Kooiman and his colleagues also makes a distinction between different levels of governance and specifically analyzes the governance triangle. In the following, our focus is therefore on interactive governance, the triangle and governance at different levels.

A starting point for Kooiman is the emphasis on the great variation in how governance takes place with actors from different parts of society who develop new institutions, arenas for interaction and collective problem solving. From this perspective, governance will be many and different institutions, and vary from one context to another. Some institutions are characterized by great complexity in terms of the participants and the problems to be solved, while others are seemingly simple, but can be challenged by the complexity of the challenge they face. The research question is to develop an analytical framework that accounts for the complexity and also enables comparison of institutions and their function.

The analytical starting point in this perspective on governance is a distinction between three societal spheres: state, market and civil society. These have ideally been analyzed as separate parts of modern societies and have had different tasks according to different principles. Kooiman's governance approach breaks with this: although governance takes place within the spheres, the governance approach implies that new arenas are developed when actors connect across the spheres.

This can be illustrated in a triangle, here the *Interactive Governance Triangle* (Figure 1.1) developed from the work by political scientists Maria Carmen Lemos and Arun Agrawal (2006) and found in various literature (Abbott and Snidal, 2010). Traditionally, governance has been linked to the upper part of the triangle with the state as coordinator and core center of power in society. The new

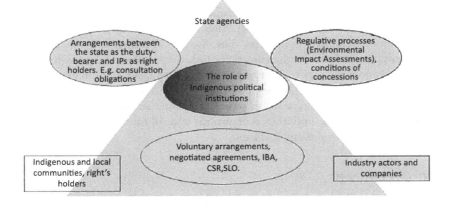

Figure 1.1 The interactive governance triangle.

concept of governance implies a change with development of new institutions further down the triangle. Attention has been given to new arenas involving actors from the market and civil society. The state might still participate, but also be absent. In a study of development in international regulation, Kenneth W. Abbott and Duncan Snidal (2010), with a background in international law and political science, have documented a phased development, first from the state to the market, and then also the establishment of institutions that are closer to the bottom of the triangle with direct relations between players in the market and civil society. Abbott and Snidal (2010) cite the Forest Stewardship Council as an example of the last phase they study and as illustrating an institution which involves actors in the lower part of the triangle. Social license to operate, where the idea is for companies to work with local communities, can also be placed in this lower part with an axis between the market and civil society.

In an Indigenous context, it is also reasonable to assume that some processes point in the direction of establishing arenas down the triangle, that is, where we identify interactions between Indigenous peoples and market actors. This is partly explained by the opportunities for direct contact between companies and civil society through, for example, social license to operate and corporate social responsibility. This leads to an activated axis toward the market. But crucial changes in the field of Indigenous peoples—new frameworks in international law through UNDRIP and ILO C169—indicate a strengthened emphasis on state-Indigenous interactions and thus a downward movement on the left side of the triangle. Central elements of international law must be understood in such a way that decisions should not be the domain of the state alone and characterized by a hierarchy of direction and management. This has been the old (governmental) notion of governance impacting Indigenous people, with institutions at the top of the triangle. Self-determination implies an expectation that decisions are moved from the state to the left corner, sometimes by establishing intermediate institutions, such as the Norwegian Sámi Parliament having governmental functions. Similarly, the requirement for consultations must be understood as a shift from hierarchy to Indigenous peoples being involved in arenas where they are regarded equal to the state.

At the same time, there is reason to maintain that the upper part of the triangle is still important for Indigenous issues. After all, the states do have a significant responsibility for human rights, and thus also for implementing the rights of Indigenous peoples. These rights can also be linked to nature and the environment, where Indigenous peoples' traditional use honors sustainable environmental management (Barelli, 2016, p. 132). Despite being viewed as an ally of big industry, the state has been an important actor in protecting nature and the environment, and this too underlines the importance of various state institutions for safeguarding Indigenous peoples' rights and inclusion.

Interactive governance

The location within the triangle clarifies the origins of participants in governance institutions; who is involved and where they come from. This has been a key aspect of some governance research. The approach by Kooiman (2003) goes a step further by placing emphasis on the actors' interaction, that is, what characterizes the interaction between them. It is the interaction of different actors that contributes to solving management challenges in a specific area, such as small-scale fisheries (Jentoft and Chuenpagdee, 2015) or industrial projects in Indigenous areas.

It may be an idealized view that the state is a hierarchical institution where decisions are issued from above, that the market is characterized by strategic behavior to maximize utility, as in negotiations, and that civil society embraces close relationships where norms govern actions and where there is equality between actors in discussions to reach agreement. Based on this approach, and considering the location in the triangle, it might be possible to identify the characteristics of interaction and decisions.

However, such an approach is problematic. It obscures, for example, that hierarchy can characterize companies and larger voluntary organizations, and that normative perceptions have a place in state institutions and company conduct. As an approach, governance is also based on a perception that it has changed and is changing, and it becomes important to examine what happens in the new arenas. Additionally, Kooiman's approach distinguishes between three main modes of governing (Jentoft and Bavinck, 2014; Kooiman and Bavinck, 2013, p. 21ff).

- Hierarchy is the form we know from the organization of states: authorities interact with individuals and groups, and develop "policy" or use management techniques to push for certain actions.
- Self-governing is linked to the ability of a collective—as a local community, an interest organization and as social movements—to govern itself without interference by other actors.
- Co-governance is characterized by equal actors who coordinate their actions sideways through coordination and cooperation. Network development is another example of co-management schemes involving stakeholders.

Levels

An important element of governance is to establish or develop institutional arrangements that provide an opportunity to solve challenges over time. These are often daily challenges and questions of a technical nature. Kooiman's approach, also enshrined in the definition of interactive governance, emphasizes that interactions may be related to questions of principles or norms that provide guidelines for daily activity and are important for the maintenance of the relevant institutions. Governance arrangements cannot just satisfy technical goals. In order to function they must have a normative basis or else they will be ineffective and lack legitimacy (Kooiman and Bavinck, 2013, p. 11).

The approach also visualizes three orders of governance. The division is a part of the conceptual framework, and is intended to capture activity related to different levels or rings of activity (Kooiman and Jentoft, 2009).

- First-order governing is related to the daily activity where the main challenge is to identify and clarify challenges as they are experienced by the actors, and in the next round look for a solution to the identified problems. For governance to work, the processes must not be purely technical: governance requires that the actors' perceived challenges emerge and that a broad search is made for possible measures.
- Second-order governing is linked to the institutional framework as formal aspects (rules, agreements and legislation) but also to the institution's norms and roles. Institutions form a core element in the context of governance, a meeting place between those who govern and those who are governed and must reflect the complexity of society and governance challenges.
- Third-order governing or meta-governance is about the overriding principles and values for governance in an area. They can be hidden and little known but can also be problematized and made the subject of problematization and change.

Discussion

The focus of this book is how actors handle challenges at a local level, the ways in which Indigenous peoples deal with major industrial projects and what opportunities they have to act and to design institutions locally. Here, the first two governance levels are central. This is where we can expect examples of Indigenous agency—what room Indigenous peoples have and how they use it to safeguard their interests.

Also, expectations at the meta-level are a key premise for examining the position of Indigenous peoples. The development in international law that Anaya (2004) describes is based on central moral premises about recognition, equal treatment and the right for a group to decide its own destiny. These are conditions that Jan Kooiman and sociologist Svein Jentoft (2009) emphasize as a part of the third level, and which in the next round need specifying. The principles must be weighed against each other and formalized through various instruments.

Seen in the context of the triangle, the state and its various institutions become the key player in balancing considerations and in linking the many levels that make up a system of Indigenous peoples' governance. It is not a given that the state acts as a coordinator and implementer. The implementation gap (Mörkenstam, 2019) indicates this. Drawing on a summary from Australia, Canada, New Zealand and the United States (CANZUS), sociologist Stephen Cornell (2019) points out that absent states lead Indigenous peoples to bypass and develop governance from below because it is most effective and provides the best opportunity to develop governance in accordance with the group's own principles. The sum is institutional diversity. The development is not coordinated, but a long-term effect can be that by taking control, Indigenous groups develop a stronger position that makes it more difficult for the state to ignore their demands in the future (Cornell, 2019, p. 27ff).

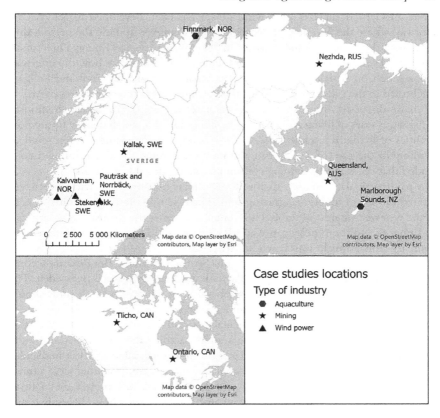

Figure 1.2 Overview of the cases and their location. Map produced by Camilla Brattland.

Case studies

The cases discussed in this book come from various geographical locations and they illuminate challenges from different industries in wind energy production, aquaculture and mining. As indicated by the map (Figure 1.2), the political and societal contexts are highly diverse. The diversity does, however, show how Indigenous peoples are affected by governmental structures and efforts to improve living conditions. When readers have reached the final chapter, we hope the cases will have broadened the understanding of Indigenous governance.

After several efforts to stimulate new projects, and genuinely supported by the public as a turn toward more renewable energy, *wind power* has become a controversial topic in Nordic countries. Some of the largest projects, with the best conditions for production of green electricity, have been located in areas of importance for reindeer herding in Norway and Sweden. The disputes over licenses have been taken to court and been debated by administrative and political bodies alike. Else

Grete Broderstad analyzes how wind power has been developed in Kalvvatnan, Norway, where the Ministry of Petroleum and Energy withdrew the permit, thus overruling the decision of the Norwegian Water Resources and Energy Directorate. Broderstad studies the argumentation of the ministry and shows how the interpretation based on Article 27 of the International Covenant on Civil and Political Rights led to the conclusion to reject the permit. Recent interpretations of Article 27 have discussed the responsibility of state authorities to secure traditional Indigenous ways of living and to avoid creating obstacles which destroy future possibilities of living in a traditional way. In Norway, the core issue is often reindeer herding. In the case of Kalvvatnan, the ministry entered the discussion and concluded that the reindeer herding community had suffered from former projects. It also paid particular attention to the development of a hydroelectric project that limited the use of traditional grazing land. The case is interesting not only in light of the renewed interest in the ICCPR but also because the Ministry of Petroleum and Energy and the Norwegian Water Resources and Energy Directorate seldom come to similar conclusions based on Article 27.

Dorothée Cambou, Per Sandström, Anna Skarin and Emma Borg examine court decisions related to the Norrbäck and Pauträsk wind energy projects in Sweden. The conflicts between wind energy developers and Sámi reindeer herding communities (*samebyar*) were handled at different court levels, and after rejection in the lower courts, the decision by the Land and Environmental Court of Appeal in 2019 authorized that the wind projects could proceed. The authors examine the argumentation by the courts, particularly related to how wind power turbines may affect reindeer husbandry. Leaning on this evaluation, the authors conclude that the courts neither serve a function as a mediator, nor solve conflicts and do not sufficiently protect the right to conduct reindeer husbandry. An important aspect to be learned from the study is that the courts have difficulty in judging the impact of the wind energy projects on reindeer herding, as there is no consensus on how to interpret the knowledge provided by the industry and the knowledge holders. A second aspect relates to the concept of sustainable development, where the courts try to meet the demands for sustainability at a meta-level but pay less attention to the fact that their interpretation undermines sustainable development of Sámi reindeer husbandry at the local level.

Aquaculture as a rising industry is new, compared to mining, but the increased investment and global growth in production has already come at a price, also for Indigenous peoples. Growth implies areal pressures, particularly on sea or water areas traditionally used for other purposes such as traditional fishing. Moreover, aquaculture may change the local economy in terms of jobs, investments and social equity. Camilla Brattland, Else Grete Broderstad and Catherine Howlett compare coastal regions in Norway and New Zealand and pay particular attention to the possibilities for Indigenous agency. The discussion and recommendations for increased agency depart from a division between structural and discursive influences. Not only are Indigenous agencies constrained rather than enabled, but the authors also argue that Indigenous rights should be strengthened, that states and private actors should be more proactive toward Indigenous peoples and that they should support capacity for participation by Indigenous organizations and coastal communities in marine development.

An example of how grassroots Indigenous peoples' organizations (*obschiny*) work directly with companies is the study by Marina Peeters Goloviznina from Russia. The study of a family-based *obschina* in the Sakha Republic illustrates how the *obschina*, assisted by the Ombudsman for Indigenous peoples' rights, managed to overcome the asymmetrical power relations with a gold mining company. The study is also instructive—even outside the Russian context—on how FPIC can be used (and misused) by companies. As the state does not define the content of FPIC, there is a risk that companies may misuse the fundamental legal meaning of the concept and deprive it of its normative value.

Another example of direct relations between Indigenous peoples and the extraction industry is the research by Horatio Sam-Aggrey in the Northwest Territories, Canada, on the relationship between the Tlichǫ people and the diamond mining industry. The case study illustrates how the Tlichǫ Agreement, an example of a comprehensive land agreement, establishes a robust legal framework that makes it possible for Indigenous peoples to take part in the management of resources on their traditional lands. This type of agreement provides clarity that benefits the industry and strengthens the role of communities in resource management and negotiations on impact benefit agreements (IBAs). In this case the Tlichǫ are active participants in the regulation of mining and in securing environmental initiatives that also include use of traditional knowledge. In addition to the implications for its relationship to the mining industry, the management of the comprehensive land agreement has strengthened the group's interaction with government agencies.

The case from Ontario discussed by Gabrielle A. Slowey is an example from an area with old treaties in Canada, and an illustration of how Indigenous rights are set aside. Mining has in general been important for economic development in Canada and Slowey argues that the protection of Indigenous rights is lost when the state continues to pursue mining to improve economic development (growth). This lopsided development has increased due to the ongoing economic crisis and illustrates the fragility of Indigenous rights. First Nations must carry the costs when government makes things easier for industry. Development of modern treaties is highly unlikely, so First Nations stand in a weak position as they lack resources to challenge the development by industry and government, and the pandemic has restricted the ability to meet and organize collectively in a meaningful way.

Catherine Howlett and Rebecca Lawrence undertake a critical analysis of Indigenous Land Use Agreements (ILUAs), the dominant agreement-making tool in Australia. They interpret agreement-making as underpinned by neoliberal logic, and although there might be positive elements for Indigenous peoples, the negative impacts outweigh the benefits. Indigenous peoples have room for agency, but it is severely limited by structural, institutional and historical realities. Agreements are not based on a real consent, but rather forced upon Indigenous peoples, and the instruments used by government and industry dispossess Indigenous peoples of resources, thus weakening their position and possibility for securing traditional culture and livelihood. The conclusion, then, is a warning for Indigenous peoples in other countries that there is "no such thing as a fair and just negotiated agreement."

In her study, Kaja Nan Gjelde-Bennett follows some of the same paths in a study of the situation in Scandinavia. The controversy over the Gállok mine in northern Sweden is the main case, analyzed from the perspective of an Indigenous paradigm versus the (dominant) neoliberal paradigm. Indigenous peoples must utilize neoliberal tools that uphold the dominant authority of the state. A solution would be to find common ground between the two paradigms where new institutions realize international Indigenous rights domestically. Gjelde-Bennett points at the proposed Nordic Sámi Convention as a possible way, as the aim is to guarantee the same rights for Sámi people living in Norway, Sweden and Finland.

One of the aims of the final chapter by Monica Tennberg, Else Grete Broderstad and Hans-Kristian Hernes, is to summarize core findings from the different cases reported in the respective chapters. The important task is to discuss findings from the governance perspective. The emphasis is on meta-governance (Kooiman and Jentoft, 2009), which focuses on normative consensus-building and clarity between different modes of governance. In contrast to recent ideas of governance, a major finding is that—despite different contexts and various arenas—the state is the most prominent actor and thus extremely important for Indigenous governance.

References

Abbott, K.W. and Snidal, D. (2010) "International regulation without international government: improving IO performance through orchestration," *The Review of International Organizations*, 5, pp. 315–344.

Anaya, S.J. (2004) *Indigenous peoples in international law* (2nd ed.). Oxford: Oxford University Press.

Barelli, M. (2016) *Seeking justice in international law: the significance and implications of the UN Declaration on the Rights of Indigenous Peoples*. London: Routledge.

Bevir, M. (2012) *Governance: a very short introduction*. Oxford: Oxford University Press.

Bradshaw, B., Fidler, C. and Wright A. (2019) "Impact and benefit agreements and northern resource governance. What we know and what we still need to figure out" in C. Southcott, F. Abele, D. Natcher and B. Parlee (eds.) *Resources and sustainable development in the Arctic*. London: Routledge, pp. 204–218.

Cascadden, M., Gunton, T. and Rutherford, M. (2021) "Best practices for Impact Benefit Agreements," *Resources Policy*, 70, 101921.

Cornell, S. (2019) "From rights to governance and back: indigenous political transformations in the CANZUS states" in W. Nikolakis, S. Cornell and H.W. Nelson (eds.) *Reclaiming Indigenous governance: reflections and insights from Australia, Canada, New Zealand, and the United States*. Tucson: The University of Arizona Press, pp. 15–37.

Dahl, J. (2012) *The Indigenous space and marginalized peoples in the United Nations*. New York: Palgrave Macmillan.

Dryzek, J.S. (1999) "Transnational Democracy", *The Journal of Political Philosophy*, 7(1), pp. 30–51.

ILO, International Labour Organization (1989) *C169 Indigenous and Tribal Peoples Convention* [Online]. Available at: https://www.ilo.org/dyn/normlex/en/f?p=NORMLEXPUB:12 100:0::NO::P12100_ILO_CODE:C169

Jentoft, S. and Bavinck, M. (2014) "Interactive governance for sustainable fisheries: dealing with legal pluralism," *Current Opinion in Environmental Sustainability*, 11, pp. 71–77.

Jentoft, S. and Chuenpagdee, R. (2015) "Assessing governability of small-scale fisheries," in S. Jentoft and R. Chuenpagdee (eds.) *Interactive governance for small-scale fisheries*. Dordrecht: Springer, pp. 17–35.

Kooiman, J. (2003) *Governing as governance*. London: Sage Publications.

Kooiman, J., Bavinck, M., Jentoft, S. and Pullin, R. (eds.) (2005) *Fish for life: interactive governance for fisheries*. Amsterdam: Amsterdam University Press.

Kooiman, J. and Bavinck, M. (2013) "Theorizing governability: the interactive governance perspective," in M. Bavinck, R. Chuenpagdee, S. Jentoft and J. Kooiman (eds.) *Governability of fisheries and aquaculture: theory and applications*. Dordrecht: Springer, pp. 9–30.

Kooiman, J. and Jentoft, S. (2009) "Meta-governance: values, norms and principles, and the making of hard choices," *Public Administration*, 87(4), pp. 818–836.

Koivurova, T., Buanes, A., Riabova, L., Didyk, V., Ejdemo, T., Poelzer, G., Taavo, P. and Lesser, P. (2015) "Social license to operate: a relevant term in Northern European mining?" *Polar Geography*, 38(3), pp. 194–227.

Lemos, M. C. and Agrawal, A. (2006) "Environmental governance," *Annual Review of Environment and Resources*, 31, pp. 297–325.

Morgera, E. (2016) "The Need for an International Legal Concept of Fair and Equitable Benefit Sharing," *The European Journal of International Law*, 27(2), pp. 353–383.

Mörkenstam, U. (2019) "Organised hypocrisy? The implementation of the international Indigenous rights regime in Sweden," *The International Journal of Human Rights*, 23(10), pp. 1718–1741.

O'Faircheallaigh, C. (2013) "Extractive industries and Indigenous peoples: a changing dynamic?" *Journal of Rural Studies*, 30, pp. 20–30.

O'Faircheallaigh, C. (2016) *Negotiations in the Indigenous world: aboriginal peoples and the extractive industry in Australia and Canada*. New York and Abingdon: Routledge.

Owen, J.R. and Kemp, D. (2017) *Extractive relations: countervailing power and the global mining industry*. London and New York: Routledge.

Piattoni, S. (2009) "Multi-level governance: a historical and conceptual analysis," *European Integration*, 32(2), pp. 163–180.

Pirsoul, N. (2019) "The deliberative deficit of prior consultation mechanisms," *Australian Journal of Political Science*, 54(2), pp. 255–271. Available at: https://doi.org/10.1080/10361 146.2019.1601681

Rosenau, J.N. (1997) *Along the domestic–foreign frontier: exploring governance in a turbulent world*. Cambridge: Cambridge University Press.

Torfing, J., Peters, B.G., Pierre, J. and Sørensen, E. (2019) *Interactive governance: advancing the paradigm*. Oxford: Oxford University Press.

Tully, J. (2004) "Recognition and dialogue: the emergence of a new field," *Critical Review of International Social and Political Philosophy*, 7(3), pp. 84–106.

Wilson, E. (2016) *What is free, prior and informed consent?* Ajluokta/Drag: Árran Lule Sami Centre.

Young, O.R. (1999) *Governance in world affairs*. Ithaca, NY: Cornell University Press.

2 International law, state compliance and wind power

Gaelpie (Kalvvatnan) and beyond

Else Grete Broderstad

Introduction

The connection between international law and the development of Sámi rights in Norway is indisputable. Because of the Alta hydropower conflict in the 1970s and early 1980s, the Sámi pushed the perception of rights into the public political consciousness by appealing to international law and human rights standards. The protests, spearheaded by the Sámi political movement in alliance with the environmental cause, paved the way for a new era in the modern history of Sámi politics. The first Sámi Rights Committee (SRC) (NOU 1984: 18) laid the foundation for new legislation followed by political institutionalization. Land rights and resource governance became the next issues to be investigated. Based on international legal standards, national authorities complied with the argumentation of the Sámi rights movement. Justified by different sources of international law (Ravna 2020), statutory revisions and consultation are in place and allow rights-holders to respond to and influence the processes applicable to them.

Still, the pressure on traditional livelihoods is increasing. Those living off the land, like the reindeer herders, face major challenges in adapting and responding to different exploitation and development projects such as mining, power lines and hydropower plants, cabin sites, roads, railroads, and—the focus of this chapter—wind power. Wind power is regarded as an important factor in Norwegian energy supplies and a part of obtaining more effective and climate-friendly energy use (White paper 2019–2020, p. 9). These projects add to already heavily exploited land areas. Underpinning all debates on industrial ventures versus Sámi traditional land use is how the governing systems intended to handle conflicting interests interact with and comprehend the challenges of those most severely affected by the exploitation activities. An illustration is the Kalvvatnan case and the argumentation of the decision-makers. I use the Norwegian name Kalvvatnan to refer to the case itself, as this is the name that appears in the case documents. The South Sámi name of the area is *Gaelpie*. A wind power plant spanning an area of some forty square kilometers was planned in the municipalities of Bindal in Nordland county and Namsskogan in Trøndelag county in mid-Norway. These areas are crucial for two reindeer herding districts, Voengelh-Njaarke and Åarjel-Njaarke. For the South Sámi as a small minority, reindeer herding is particularly significant for maintaining language, identity and culture (Fjellheim 1991, 2020).

DOI: 10.4324/9781003131274-2

I accentuate the full spectrum of international legal obligations which protect Indigenous rights. The global standards for the relationship between the state and Indigenous peoples are authorized by the United Nations Declaration on the Rights of Indigenous Peoples (UNDRIP). Still, my focus is on Article 27 of the International Covenant on Civil and Political Rights (ICCPR), weighed by legal experts in land rights conflicts (NOU 2007a, 2007b; Ulfstein 2013). I will especially focus on the assessment of the Ministry of Petroleum and Energy (OED) of this provision, because the wind power project was abandoned. Article 27 of the ICCPR runs as follows:

> In those States in which ethnic, religious or linguistic minorities exist, persons belonging to such minorities shall not be denied the right, in community with the other members of their group, to enjoy their own culture, to profess and practice their own religion, or to use their own language.

Three reasons guide the choice of this angle. First, the application of Article 27 stretches back to the 1980s, when it became the benchmark against which state behavior toward the Sámi was assessed. This underlined the state's responsibility to secure the material basis of Sámi culture (see the next section). Second, in 1999, the ICCPR was given precedence over internal legislation due to the adoption of the Human Rights Act (1999) incorporating the human rights conventions. Third, a review of the Kalvvatnan case is interesting because Article 27 was the benchmark in the assessment and argumentation leading to the Ministry's decision to halt the wind power plans. Regarding the two first mentioned reasons, comprehensive legal literature provides insights into Article 27 in terms of the state's duty to secure the Sámi people's cultural protection and the implications of this protection (cf. Bull 2018; NOU 1984: 18; NOU 2007a, 2007b; NIM 2016; Ravna 2014; Smith 2014; Smith and Bull 2015; Ulfstein 2013; Åhrén 2016). While I emphasize Article 27 in the case of Kalvvatnan, which has been less researched, I am aware of the OED assessments of this provision in other wind power cases leading to opposite results (see section 'Selected reference cases'). Each case deserves an in-depth analysis, but I will limit my focus to Kalvvatnan and will only mention the others as selected reference cases in the empirical and discussion sections to the extent that they might illuminate some core aspects of state compliance with procedural and material requirements of international law.

My starting point is the concept of state compliance and how international human rights obligations constitute requirements on states. These include material requirements as legal rules framing and determining the content of decisions. Procedural requirements contain demands on how cases should be processed. By illuminating the reasoning of the ministerial decision based on standards anchored in Indigenous, minority and general human rights norms, in particular Article 27, I aim to contribute to the discussion of state compliance with international law applicable to Indigenous peoples. The question is: *given the formal significance of international law in the Sámi–state relationship, how can the reasoning in concrete decisions such as that of Kalvvatnan illuminate state comprehension of procedural and material requirements of human rights norms?* Clarifying the decision and its reasoning could turn out useful in a discussion on relevant concerns, determining when similar issues should be treated equally, and different issues differently. In other conflicts over wind power, reindeer herding communities who are affected frequently refer to Kalvvatnan, claiming that the same reasoning should apply to them.

I fully acknowledge the efforts of, for example, Sámi youth (Ságat 2017; Lund, Gaup and Somby, 2020, p. 55) and other actions taken by the reindeer herding communities against this project (NRK Sápmi 2014, 2015). But here I will concentrate on reviewing the decision of the Ministry, as well as drawing on the consultation minutes and other official case documents revealing argumentative claims and premises of an institutional character. I will present the content of the ministerial decision before analyzing the contents through the lens of compliance indicators. This is not a legal study and as a political scientist, I provide some conceptual insights. The choice of emphasizing a single case and introducing single aspects of similar cases might be premature, as a larger empirical material could reveal patterns and connections of better explanatory power. Still, insight into the Kalvvatnan proceedings will illuminate state comprehension of core principles of procedural and material requirements of international legal standards. While the state remains the primary duty-bearer of human rights obligations (NIM 2019), the growing recognition of Indigenous rights nationally and internationally changes the legal and political context of business enterprises (cf. O'Faircheallaigh 2013), as the governance triangle in analytical terms depicts (Hernes, Broderstad and Tennberg, in this volume), a point revisited at the end of this chapter.

In the next section, I briefly discuss the role of Article 27 of the ICCPR in the Sámi rights development in Norway. In the third section, I introduce the concept of compliance and the distinction between structure, process and outcome as indicators of state compliance, with a primary focus on the two latter aspects. Thereafter, I present and highlight the ministerial explanation in the case. To shed light on state behavior beyond Kalvvatnan, I take a brief glance at selected aspects of the reference cases. Leaning on the legal review of the second Sámi Rights Committee (SRCII), the empirical review is then discussed with the compliance indicators serving as an analytical framework, before conclusion.

Background: interpretations of international law

The Norwegian Sámi Parliament, established in 1989, has become an institution of significant political influence and symbolic strength. Norway has a reputation as a human rights guardian, has ratified a series of human rights instruments and was the first country, in 1990, to ratify the International Labour Organization (ILO) Convention No. 169 on Indigenous and Tribal Peoples in Independent Countries (ILO 169, International Labour Organization Convention 1989). Several domestic acts have been passed to uphold human rights obligations. When adopting the Sámi Act in 1987 and the constitutional amendment in 1988, the Norwegian Parliament thoroughly debated and acknowledged the impacts of the long-lasting official assimilation policy by the Norwegian state. The Sámi Rights Committee (NOU 1984: 18) raised arguments of "being conscious about the past as a condition for the future," "the authorities' most solemn commitment," and "a barrier towards policies impairing Sámi culture." These had a prominent role as a moral motivation for the legislature (see Semb 2001; Broderstad 2008). In 1999 Norway adopted the Human Rights Act incorporating the ICCPR with precedence over internal legislation. The dualistic structure of the Norwegian legal system implies that "international law must be implemented through an act of the Parliament to become national law" (Ravna 2020, p. 149). In 2007, Norway voted in favor of UNDRIP in the UN General Assembly.

The Sámi Rights Committee's interpretation of Article 27 implies an active contribution by the state to develop Sámi culture, while embracing the material aspects of a minority culture. The committee based their proposal for a constitutional provision on state responsibility to secure Sámi language, culture and societal life on this interpretation of Article 27 (NOU 1984: 18, pp. 438, 441). Their reading was followed by the Norwegian Parliament and figured prominently during the Sámi institutionalization process in the late 1980s. The Sámi Rights Committee emphasized the duty of the state to develop the material basis of Sámi culture, which has been the prevailing view ever since. According to international relations and law expert Beth A. Simmons (2009, p. 5), treaties reflect politics, but they also "shape political behaviour, setting the stage for new political alliances, empowering new political actors, and heightening public scrutiny," as shown by the changes and learning processes of the 1980s in the aftermath of the Alta conflict.

Article 27 of ICCPR is the cornerstone of international minority law (Barten 2015). In general terms the Human Rights Committee (HRC) has emphasized the applicability of this Article with respect to Indigenous peoples, arguing "that indigenous peoples can constitute minorities and that cultural rights can be of special relevance for indigenous peoples" (Barten 2015, p. 50). As a UN treaty-based body, the committee monitors the implementation of the ICCPR by the state parties, examines state reports, considers inter-state complaints and examines individual complaints of alleged violations by state parties (UN Human Rights website n.d.). The applicability of Article 27 has been reinforced by the committee's recognition of interdependence between Article 1 on self-determination and Article 27 (Scheinin 2004, p. 9; Åhrén 2016), especially in decision-making processes concerning the use of land and natural resources (cf. NOU 2007a, p. 186). According to the second Sámi Rights Committee, the fact that HRC examines individual complaints increases clarity and transparency. The HRC opinions, statements and decisions on individual complaints have a significant impact by virtue of being the source of law for comprehending international law (NOU 2007a, p. 177). This Sámi Rights Committee was appointed by the government in 2001 to report on questions relating to the Sámi population's legal position as regards the right to, and disposition and use of, land and water in traditional Sámi areas south of Finnmark (NOU 2007a, point 2.5.1).

State compliance

In scrutinizing the ministerial reasoning, the point of departure is the concept of compliance, and the question to what extent the government follows through to implement standards of international law. Political scientist Harold K. Jacobson and legal scholar Edith B. Weiss (1995, p. 123), maintain that compliance can "refer to whether countries in fact adhere to the provisions of the accord and to the implementing measures they have instituted," while scholar in international law Benedict Kingsbury points out that compliance "derives meaning and utility from theories, so that different theories lead to significantly different notions of what is meant by 'compliance.'" It signifies more than simple correspondence of behavior with legal rules (Kingsbury 1998, p. 346).

A debate on how to comprehend "compliance" in relation to different theoretical premises is beyond the scope of this chapter. Compliance is not an either/or issue (Jacobson 1998, p. 570). Rather the issue is to determine to what extent nations follow their obligations. Or, as the former Senior Adviser on Human Rights, International Labour Office, Lee Swepston (2020, p. 114) states:

> simple causal relationship is relatively rare either in the ILO or the UN, or in other international process can also be severely underestimated. International pressure is rarely the only reason governments take action, and few governments wish to admit that they are "giving in" to this kind of pressure, but it is an important element in decisions to act in a way that conforms to their international obligations.

Discussing the effects of the supervisory system of ILO 169, Swepston (2020, p. 114) emphasizes that

> international supervision is vital to bringing situations to light, but ... it is almost never the sole reason for a specific national outcome. International standards and their supervision can stimulate action by calling attention to problems that are then taken up in the national political or judicial process, and by providing a standard against which solutions can be measured.

How these standards are made topical is in the limelight here. As will be pointed out, the ministerial decision on Kalvvatnan concerns compliance both as substance and procedure. Substance compliance deals with whether states comply with norms and rules of international law, such as the material limit for interventions on lands on which Indigenous peoples depend. A report released by the Norwegian Water Resources and Energy Directorate (Dalen and Villaruel 2019, p. 7) discusses how Sámi interests are protected by international law in the context of wind power. When the directorate assesses whether to permit hydropower or power lines, they must assess if and how domestic or international law delimit their decisions. This follows, among other things, from the protection enshrined in Article 27 and the constitutional section §108 on state authorities' duty to enable Sámi people to preserve and develop their language, culture and way of life.

As the name suggests, procedural compliance is concerned with obligations of a procedural nature, such as adherence to rules of consultations and proceedings. Procedural requirements demand how cases should be processed, for example, according to the Public Administration Act (1970), the Energy Act (1990), the

Planning and Building Act (2008) and the Reindeer Husbandry Act (2007). In accordance with the Planning and Building Act (§ 5–4), the Sámi Parliament has the right to object to plans related to issues of significant importance to Sámi culture or conduct of commercial activities. Regulations of international and domestic law protecting Sámi interests also imply intensified demands on regular case proceedings (such as the annual report of the Parliamentary Ombudsman 2009).

ILO 169 is Norway's most explicit commitment to Indigenous rights (ILO 2016, p. 13), and

> has significance in accordance with the presumption principle in Norwegian law, which states that Norwegian law presumes to comply with international law, particularly in cases where several interpretative alternatives to Norwegian law exist.
>
> (Ravna 2020, p. 150)

The comprehensive interpretive material and concretization by the ILO bodies of enforcement on the consultations and participation provisions make these provisions principal (NOU 2007b, p. 829). The consultation agreement of 2005 between the Norwegian government and the Sámi Parliament (Consultation agreement 2005) can be understood in light of the procedural nature of compliance. The same can be said about the separate consultation agreement of 2009 between the Sámi Parliament and NVE (Consultation agreement 2009). An ILO report on experiences from Norway (ILO 2016, p. 5) accentuates that

> If indigenous peoples' rights, concerns and aspirations are reflected in legislation and broader policies, it will likely be easier to reach agreement and consent on specific measures or projects affecting their lands and territories.

The Storting—the national parliament—has in June 2021 adopted a bill on changes in the Sámi Act (Draft Resolution and Bill 2020–2021), making the obligations and rights of consultation statutory.

The institutionalized process promotes the involvement of the Sámi Parliament in state decision-making processes. Through formalized consultation arrangements the Sámi Parliament is consulted at different stages of, for instance, wind power proceedings. The right to be consulted also applies to "local Sámi communities and/or specific Sámi entities or interests that may be directly affected by legislation or administrative measures" (Procedures for consultations 2005, point 9). A ministry guide on consultations further emphasizes the reindeer herders' right to consultations (Ministry of Labour and Inclusion 2006).

To shed light on state performance, I draw on the categorization of human rights indicators as structural, process and outcome indicators assessing state compliance (De Beco 2008, p. 42; Indigenous Navigator website; NIM 2020; OHCHR 2012). Simultaneously, I emphasize that by analyzing the government's assessment, I seek to establish how the government in qualitative terms comprehends their human rights obligation. This study does therefore not apply the concept of indicators as tools of a quantitative assessment of human rights compliance. Human rights law expert

Gauthier De Beco (2008, p. 42) distinguishes between benchmarks as the targeted performance and indicators as the actual performance and discusses a framework of measuring the extent to which duty-bearers fulfill their obligations and right-holders enjoy their rights. *Structural indicators* assess whether states have committed themselves to protecting human rights, for example through the ratification of human rights treaties and by integrating them in domestic legislation, and the formation of non-judicial institutions to monitor the implementation of human rights. *Process indicators* evaluate state policies on implementation, efforts that, for example, enable people to participate in processes of implementing human rights. *Outcome indicators* assess state performance and evaluate de facto compliance with human rights treaties, focusing more on the results than on the efforts themselves. As De Beco (2008, p. 45) points out, these indicators are only useful if combined with each other, for they are complementary and interdependent.

"While structural indicators evaluate the commitments undertaken by states, process and outcome indicators assess their abidance by them" (De Beco 2008, p. 45). For instance, a decision to protect reindeer herding pastures depends on state legislation and the policy of the state toward reindeer herding and reindeer herders' access to participation in decision-making. State behavior can also be measured from the premise of human rights violation and enjoyment. One stresses the state's failure to comply with human rights, the other refers to the extent to which rights-holders enjoy their rights (De Beco 2008, p. 31). Structural and procedural indicators are generally applicable, outcome indicators are specific. This attempt to discuss the implementation of international legal standards and, in particular, Article 27 through the lens of such categorization is revisited in the discussion section.

The case of Kalvvatnan power plant and the ministerial decision

Background and overall review

In March 2014, the Water Resources and Energy Directorate (NVE) licensed the company Fred Olsen Renewables AS, which in 2006 had notified the directorate about their plan to build the Kalvvatnan wind power plant of up to 225 megawatts in the municipalities of Bindal and Namsskogan. In all larger wind power cases, the directorate receives notification as an early warning of a planned project, promoted pursuant to the Planning and Building Act's regulations on impact assessments. According to the processing of licenses in wind power development, all planned endeavors of more than 10 MW trigger an impact assessment program before application to the NVE. The directorate's licensing decision can be appealed to the Ministry of Petroleum and Energy, whose decision is final. In most cases the Ministry has upheld the NVE decisions, but "it rejects more licenses than the NVE does, indicating a stricter practice" (Gulbrandsen et al. 2021, p. 7). Before the licensee can develop the project, the directorate must approve a detailed plan of environment, transport and construction, a so-called MTA plan (Miljø-, transport og anleggsplan) (Dalen and Villaruel 2019; NVE 2020).

The NVE received the licensing application in October 2011, with different thematic impact assessments as a basis, including one on reindeer herding. Public hearings and meetings followed. In December 2011, the directorate consulted the affected reindeer herding districts (NVE 2014). The Directorate for Cultural Heritage, the then area boards of the reindeer herding districts, the County Governor of Nordland and the Sámi Parliament raised objections against the project. The Sámi Parliament was concerned about the total land use situation and pointed out that the critical level for sustainable reindeer herding had been reached long ago due to earlier encroachments. The area boards addressed the value of the areas for the two reindeer herding districts, disagreeing with the evaluation and conclusions of the impact assessment which, according to the reindeer herding districts, had a host of major shortcomings. Still, these concerns did not change the decision of the NVE.

Consultations

At a meeting on December 5, 2012 (Consultation Protocol 2012), the Sámi Parliament and the Water Resources and Energy Directorate consulted on seven applications for wind power plants in the counties of Nordland and Finnmark. One of these was Kalvvatnan. The Sámi Parliament objected to all the plans and decided not to withdraw their objections. The parties disagreed on what the appropriate procedures implied. The Sámi Parliament considered that the obligation of the state to consult had not been fulfilled; no real consultations had been conducted; thus, additional consultations were necessary. The NVE made it known that the Sámi Parliament's request with regard to receiving draft decisions had to be principally clarified between the Sámi Parliament and NVE and, until such clarification, presenting draft decisions as a basis for consultations in the wind power cases was not an issue. In these circumstances, the NVE declared, it would be impossible to reach an agreement (NVE 2014, p. 12).

After the NVE had licensed the company in 2014, the Sámi Parliament notified the Ministry about the consultation procedures and not having received information about the directorate's considerations and draft decision. Non-compliance by the NVE with the obligation to consult had, according to the Sámi Parliament, influenced the decision which could not be regarded as legal. The Sámi Parliament hence urged the Ministry to contact affected Sámi interests to inquire about consultations. If desirable, the Sámi Parliament (2014) could then participate as an observer. As the party responsible for handling the complaints, the Ministry held an appeal inspection and meetings with different bodies—the districts, youth and interest organizations—between June 2015 and August 2016. The two reindeer herding districts were consulted in January 2016. In October, the Ministry and the Sámi Parliament met for consultation (Consultation Minutes 2016), and on November 11, 2016, the Ministry decided not to license the wind power plant in Kalvvatnan.

Selected reference cases

By applying the Ministry's final decisions on Fosen in 2013, Hammerfest and Fálesrášša, in 2015 and Øyfjellet in 2016 as reference cases, the vagueness in the state's assessment of the protection of Article 27 reveals itself. In conjunction with the Fosen project, a report (Ulfstein 2013) on the principles of international law was produced, constituting a basis for future cases. In 2016, five days after Kalvvatnan, the Ministry confirmed a directorate decision to license another wind power venture in the South Sámi area, namely the Øyfjellet wind park of 330 MW (where the volume was later increased to 400 MW) affecting reindeer herding in Jillen Njaarke in Vefsn municipality in the south of Nordland county. The closeness in time between Kalvvatnan and Øyfjellet is interesting with the considerations on Kalvvatnan fresh in minds. The directorate's decision in November 2014 to license Eolus Vind AS in the Øyfjellet case, had been appealed by the reindeer herders and collaborating environmental interests in Nordland (OED 2016b). The result of this case followed in the steps of the Fosen case in Nord Trøndelag. In June 2010 the directorate had licensed Europe's largest land-based wind power complex, including Storheia of 288 MW. The Ministry of Petroleum and Energy found that neither Øyfjellet nor the Fosen plant conflicted Article 27. Opposite to the cases of Fosen and Øyfjellet are the Hammerfest and Fálesráššá cases from 2015, in Finnmark in the North Sámi area. In Hammerfest, motivated by reindeer herding concerns, the directorate rejected the wind power plan of 110 MW, leading to an appeal by the proponent and Hammerfest municipality, but the NVE decision was endorsed by the Ministry (OED 2015a). In the Fálesráššá case in Kvalsund municipality, the NVE licensed a 180 MW plant, but appeals by the reindeer herders and environmental interests made the Ministry re-examine the directorate's decision and reject the application due to concerns for the reindeer herding communities (OED 2015b). Hammerfest, Fálesráššá and Kalvvatnan stand out as cases where the Ministry refused to license the wind power plans due to concerns for the reindeer herding communities, while Fosen and Øyfjellet could proceed.

Worth mentioning is also an aspect of procedural character. At a consultation meeting in October 2016, the Ministry and the Sámi Parliament agreed on the terms of the license granted to Eolus Vind AS in the Øyfjellet case:

> The licensee has to arrange for entering into an agreement with the Jillen-Njaarke reindeer herding district on the proposed mitigation measures for reindeer herding in the construction and operation phases. The proposal must among others secure access to winter pastures in the northwest with mitigation measures linked to the migratory trails through the planned area. Measures of securing migration to and from winter pastures in the northwest shall be submitted to the County Governor for assessment in accordance with the Reindeer Herding Act. Proposals of mitigation measures must be included in the detailed plan of measures to be taken, cf. term 13. The detailed plan must be approved by the NVE. If consent between the licensee and the reindeer herding district is unobtainable, the NVE must consult the district before a detailed plan can be approved (my translation).
>
> (Consultation Minutes 2016)

The Ministry emphasized these conditions as decisive in the licensing of the Øyfjellet wind power plant (OED 2016b, see term 16 of the conclusion). In December 2019, the NVE approved the detailed environment, transport and construction plan, and decided in January/February 2020 not to postpone the implementation of the decision despite appeals from, among others, the reindeer herders. The parties disagreed on the implementation of the proposed mitigation measures out of concerns for reindeer herding. However, the directorate decided on further mitigation measures, demanding that the licensee halt the construction of a road for a month during migration and provide resources for gathering and migration. The company requested to have the decision delayed, which the Ministry approved in April 2020 (NVE 2020), resulting in what was experienced as a chaotic and stressful spring migration by the reindeer herders in Jillen Njaarke. While the case was significantly more complex than described here, I include this point as an illustration of a gap between state behavior and state politics, revisited in the discussion section.

Considering the case of Kalvvatnan

The following presentation draws on the review and decision by the Ministry of Petroleum and Energy in the Kalvvatnan case on November 11, 2016 (OED 2016a). The Ministry declares that the advantages and disadvantages of the case must be considered, and one should also assess the energy needs, long- and short-term environmental consequences and the Sámi interests. The inconvenience and damage caused to Sámi interests must be assessed in light of minority protection stipulated by international law and considered against the limits of what is and is not permissible (OED 2016a, point 3.1).

The interpretation is based on the principles of international law with reference to the mentioned expert report by legal scholar Geir Ulfstein on Sámi international legal rights in cases of damage to nature (Ulfstein 2013). The Ministry states: "Sámi reindeer herding is protected by general Sámi Indigenous and minority rights" (OED 2016a, p. 3). An appendix elaborates the Ministry's assessment of the protection of Indigenous peoples by international law, mentions the consultation obligations of ILO 169 and emphasizes Article 27 of ICCPR. The Article plays a significant role as it constitutes protection against measures denying or constricting the practice of Sámi reindeer herding. Whether such measures will be in defiance of Article 27 depends, according to the Ministry, on a concrete assessment of whether the possibilities to exercise Sámi culture are violated or denied. Therefore, "a measure that does not imply a total denial of cultural practice will still violate article 27, if it considerably restricts Sámi cultural practice" (OED 2016a, p. 3).

Discussing the implications of such a violation or denial, the Ministry leans on the work of the UN Human Rights Committee. Crucially, in assessing whether Sámi cultural rights and practice have been violated or denied, one should consider the assembled impact of past and present measures and whether these combined deny Sámi reindeer herders the right to exercise Sámi culture. The assessment of any violation of Article 27 should also consider whether the reindeer owners have been consulted and whether negative impacts have been mitigated. Thus, the Ministry declares: "An infringement of the right to cultural exercise will represent a breach of article 27. In that case it is not possible to grant a license in accordance with § 3–1 of the Energy Act" (OED 2016a, p. 3).

The Ministry goes on examine reindeer herding use of the areas involved, including pastures, calving and calf marking areas, reindeer trails, migration routes, gathering and slaughtering areas during different seasons. The area of the planned wind power plant is used by the two reindeer herding districts separately and jointly depending on the season. The districts are already affected by other energy projects and infrastructure, as was made clear by representatives of reindeer herding at consultations on a hydropower project in 2012. According to them, the critical level of new disturbance had been reached (OED 2016a, pp. 4–5). After the construction of the Tosbotn hydropower project, the Ministry realized the necessity to safeguard the remaining resource basis for reindeer herding and, according to a royal decree (Royal decree of June 22, 2012 on the hydro power project of Tosbotn, in OED, in 2016a), licensing authorities must consider the total burden on reindeer herding when future energy development applications are weighed. As a condition for licensing the Tosbotn project, the assessment of the cumulative impact on reindeer herding concludes that

[l]arge grazing areas along the original lakes have been lost. In the process, several tracks previously used by reindeer herding have been redrawn due to changed water conditions in the lakes and several rivers in the area (…). In general terms the hydropower developments have complicated the drive and use along the river systems, and have led to, among other things, increased pressure on the neighboring districts. (…)

There are also a series of other disturbances negative for the district … [such as] the E6 highway and the railroad, the Tosen road, cabin building, forestry and several power stations (my translation).
(Nybakk et al. 2013 in OED 2016a, p. 5)

The section on wind power and impacts on reindeer herding (OED 2016a, pp. 5–7) explores the use of the areas in reindeer herding, the value of the different areas and the impacts on reindeer herding by the construction and operating stages, with reference to the impact assessment. The reindeer herders argue against the assessed value of the grazing areas (cf. Eftestøl et al. 2011), which they consider too low. The proposed mitigation measures imply increased disturbance and reduced use of pastures and airing, marking and calving areas. The NVE concludes that there is no certainty of the actual impacts, which they claim would primarily appear during the construction phase. Also, while the Ministry notes that the research results on the impacts of wind power are divergent, they refer to a Swedish report, published after NVE´s decision, on three studies of wind power showing negative impact on reindeer and reindeer herding (Skarin et al. 2016). These studies report reduced and disturbed grazing within a three- to four-kilometer distance from the wind power plants. The Swedish studies also show that the reindeer prefer habitats where they cannot see the turbines.

The review further deals with migration routes, the construction phase, access road and South Sámi culture. Migration routes have special protection in § 22 of the Reindeer Husbandry Act. This is duly noted by the Ministry, which foresees considerable challenges during the construction phase in utilizing the northern planning area as migratory trails. Permanent challenges of this kind would signify defiance of the law. Construction should, according to the impact assessment, be carried out when the reindeer are elsewhere—ideally, in winter—but this option is unrealistic. Thus, the assessment proposes that the construction begin no earlier than June 15, and preferably not before July 1. Also, avoiding co-occurring construction would make one of the high grounds available for reindeer in the summer. The Ministry assumes that these restrictions and adjustments will complicate the construction phase, which would stretch across several snowless seasons and entail increased expenses. "Even with considerable restrictions on construction, the Ministry believes that the construction will present large challenges to reindeer herding" (OED 2016a, p. 9). Road alternatives for the construction and operation are further assessed, especially as the directorate and the Ministry disagree on the best option for the access road. This is related to additional use and disturbance, and to an increasing number of cabins, hunters and others in the area. The directorate emphasizes the ripple effects of their options for one of the affected municipalities. The Ministry, however, regards these options as unviable, as they have not been sufficiently shown to be socially rational in the sense laid down in the Energy Act, and the alternative favored by the directorate does not accord with the principle of proportionality in international law (OED 2016a, p. 12).

The Ministry also discusses the significance of reindeer herding for South Sámi culture and language. South Sámi is on the UNESCO list of seriously endangered languages, and the Ministry concludes that reindeer herding is an important carrier of South Sámi language and culture. In the process of assessing and deciding whether to license wind power, the Ministry finds—considering wind power's advantages and disadvantages—that reindeer herding is in an exceptional position. The Ministry concludes that a new wind power plant, with all its ramifications and consequences, runs the risk of threatening the maintenance of current levels of reindeer herding in the area.

In summary, the Ministry makes it clear that if an energy proposal is in breach of Article 27, it is not possible to license such an endeavor as set out in the Energy Act § 3–1. The assessment of an endeavor includes a judgment of whether the assembled impact from wind power and previous infrastructure would violate or deny reindeer herders the right to exercise their culture. As a second point, the Ministry estimates the value of the area and how it is used by reindeer herding. The implications of earlier encroachments are similarly assessed. In discussing the impact of wind power on reindeer herding, the Ministry of Petroleum and Energy lists the negative effects identified in the impact assessment and juxtaposes these with the judgment of the reindeer herding districts and administration, which in the hearing process disagree with the consultants' judgment of low-level impacts (OED 2016a, p. 6). While the Ministry refer to the research results as ambiguous, they mention research findings which show that reindeer avoid wind turbines. The Ministry also finds it hard to locate alternative migration routes if existing trails have to be closed. The impact assessment's recommendation of carrying out construction work when no animals are present will—according to the Ministry—lead to restrictions and adjustments that will raise the costs and extend the building period. And finally, reindeer herding is acknowledged as an important carrier of South Sámi language and culture (see also Olsen 2019, pp. 15–18; Anti 2020, pp. 28–31).

Discussion

To answer the question of how Kalvvatnan can illuminate state comprehension of compliance within the norms and rules of international law, I will discuss the emphases and reasoning of the government's decision through the categories of structure, process and outcome.

De jure compliance

As mentioned in the background section, Norway has a "global reputation as upholders of international law and conventions" (Minnerup and Solberg 2011, p. 4). Legislation and institutionalized policies on Sámi issues as Indigenous matters constitute *de jure* compliance. But as the second Sámi Rights Committee (NOU 2007a, p. 175) points out, incorporated provisions of conventions do not necessarily have much of a practical impact until they are asserted as the basis for concrete rights and duties or as the foundation for the authorities' practice. Article 27 has a long history of underpinning the consolidation of Sámi rights. The limitations on what the state can do or permit others to do without violating the provision were emphasized by the Ministry in Kalvvatnan and substantiated by international and national law. The preparatory work of the Reindeer Husbandry Act underscores the significance of Article 27 for safeguarding the material basis of Sámi culture and the legal protection of reindeer herding, which together with ILO 169 determine which encroachments on reindeer herding are permissible (Draft Resolution and Bill 2006–2007, p. 15). Norway has ratified and is committed to ILO 169 as the most explicit expression of the recognition of the Sámi as an Indigenous people. It contains several provisions concretizing the obligation of the state to consult Indigenous peoples. Consultations should follow appropriate procedures in a climate of mutual trust; the government should recognize Indigenous representative organizations; all relevant information to the Indigenous party should be ensured; both parties should endeavor to reach an agreement and avoid unjustified delays, but also secure sufficient time to allow Indigenous peoples to engage in decision-making and comply with the concluded agreements and implement them in good faith (ILO 2016, p. 4).

Still, structural indicators—such as the recognition of Indigenous rights in national legislation (see Indigenous Navigator 2020, p. 9), material thresholds or principles of effective participation—cannot alone evaluate state compliance with human rights treaties, but they do serve as benchmarks against which state behavior can be assessed, and also relate to developing other types of human rights indicators (De Beco 2008, p. 42).

De facto compliance

Process indicators measure de facto compliance with human rights treaties and can evaluate enabling aspects of participation (De Beco 2008, pp. 43–44), such as the conditions of consultations, including information sharing and right-holders' involvement and knowledge base in impact assessments. Consultations undertaken by the state crucially help assess whether the procedural aspect of Article 27 has been violated, as pointed out by the OED in Kalvvatnan and other cases. The second Sámi Rights Committee highlights the Human Rights Committee practice on Article 27: in their assessment of compliance or not, the committee has in several cases emphasized effective and active participation by the affected Indigenous groups (NOU 2007a, 2007b, pp. 206, 859). Still, Article 27 is less specific and the obligation to consult more indirect than it is in ILO 169 (NOU 2007b, p. 824).

In the Kalvvatnan consultations between the Sámi Parliament and the Norwegian Water Resources and Energy Directorate in 2012, the directorate refused to provide their considerations on the draft decision to the Sámi Parliament. The NVE thus declined to present their reasoning and draft decisions as a basis for consultations, which the Sámi Parliament found did not fulfill the obligation of securing all relevant information. This the parliament saw as not complying with the obligation to consult. Another round of consultations took place in 2016, referred to by the Ministry in their account of the case (OED 2016a, p. 4). The reindeer herders described their use of the Kalvvatnan area, and the Ministry referred to earlier consultations on hydropower in 2012, where an assessment of the cumulative impact situation (Nybakk et al. 2013, in OED 2016a, p. 5) became a condition for licensing the project. In the consultation between the Sámi Parliament and the Ministry in October 2016, the parliament was satisfied that these premises were honored in the Kalvvatnan considerations (Consultation Minutes 2016).

This conduct can be viewed in light of the consultation standards of complying with concluded agreements and implementing them in good faith. A different approach is apparent in the Ministry's decision in April 2020 on Øyfjellet, which delayed the effectuation of the NVE decision on mitigation measures during the spring migration (NVE 2020). Earlier on, the Ministry and the Sámi Parliament had agreed on the terms of the license (Consultation Minutes 2016), requesting the licensee to reach an agreement with the reindeer herding districts on mitigation measures of impacted migration routes during the construction and operation phase of the project. These terms were also stressed in the Ministry's own decision on Øyfjellet (OED 2016b).

The Kalvvatnan case reveals that NVE and OED held different views on how one should deal with the uncertainty over the impacts wind power has on reindeer herding. The directorate's uncertainty about the actual impact led to a claim that the reindeer could use the area in the future, but land use would be temporarily reduced (OED 2016a, p. 7). The Ministry recognized the uncertainty and lack of clear research results and referred to Swedish specialist reports with different findings from previous studies. In the absence of secure knowledge, the Ministry took a precautionary approach to the uncertainties over the scope of damages and disadvantages of the project. In the Hammerfest case (as in Fálesráššá) the previous year, the Ministry declared that reindeer herding would be too negatively impacted by a windmill plant (OED 2015a, 2015b), while in Fosen in 2013, the Ministry pointed out that the plaintiffs disagreed with the impact assessment on the effect on pastures due to the encroachments, but assessed the case on the premise that wind power and power lines would not prevent the use of the area as grazing land also in the operation phase (OED 2013, p. 99). The experiences of inadequate impact assessments have been systematically raised by Sámi right-holders citing the lack of a holistic approach in impact assessments, the need to carry out these assessments at an earlier stage of the project circle, and the need for mechanisms that allow Indigenous peoples to question evaluations, methods and findings of impact assessments (NIM 2019, p. 4). However, a White Paper (White paper 2019–2020) indicates improved involvement of reindeer herding and other Sámi actors in the licensing process (p. 39), and suggests enhancing reindeer herders' involvement in the impact assessment procedures including discussing who should carry out the assessments with the reindeer herders (p. 40).

While consultations and active participation are necessary conditions for compliance with Article 27 and ILO 169, they are not necessarily adequate. An evaluation of whether consultations have been carried out or not will say something about how the state enables people to participate. However, the state cannot consult itself away from the more absolute demands for cultural protection (NOU 2007a, pp. 208–209), a point which leads us to a discussion on the results of state compliance.

Results of state compliance

Outcome indicators focus on the results of efforts taken by the state, and "some outcome indicators are directly provided for by human rights treaty provisions" (De Beco 2008, p. 44). Article 27 specifies such an obligation, directed toward results of state efforts and omissions (NOU 2007b, p. 1090). The provision assigns the state a result responsibility to protect the viability of Sámi culture (NIM 2016, p. 23). In the Kalvvatnan case, the government's decision can be said to comply with the material requirements of Article 27, or with the standards against which the results can be assessed. Material requirements entail the need to assess the cumulative impacts of historical inequities and recent development to decide whether an intervention collides with Article 27. If the rights to exercise Sámi culture are dishonored, the state violates these very rights—a concern that is in fact raised by the Ministry. Their examination of the impacts of existing projects and infrastructure in Kalvvatnan and the stress this has caused reindeer herding, leads the Ministry to conclude that the sum of established disturbance and that from a new wind power plant would reduce the levels of current reindeer herding. The ministerial decision in this case accounts for both the standards of "no margin of appreciation in assessing the range of article 27" and "the need to assess the sum of historical inequities and recent development."

Another aspect is that states have no "margin of appreciation" in assessing the range of Article 27 in concrete cases. In Ilmari Länsman v. Finland it is clearly stated that

> A State may understandably wish to encourage development or allow economic activity by enterprises. The scope of its freedom to do so is not to be assessed by reference to a margin of appreciation, but by reference to the obligations it has undertaken in Article 27.
>
> (Communication No. 511/1992 para. 9.4. in NOU 2007a, pp. 195–196)

Measures constituting a denial of cultural rights are not permissible except in times of national crisis (cf. NOU 2007a, p. 195). In its assessment of Kalvvatnan, the Ministry of Petroleum and Energy acknowledges, among other things, that international law establishes the absolute limit on what kind of endeavors can be allowed. Businesses overstep the mark if their plans considerably violate the possibility of cultural practice. Thus, compliance with Article 27 in a specific case such as Kalvvatnan is more of an either/or issue: the provision sets an absolute limit (to violation or denial of the rights), and if the planned project crosses that limit, it is not allowed to proceed. The provision does not, however, provide absolute protection (NIM 2016, p. 12; Dalen and Villaruel 2019, p. 15; NOU 2007a, pp. 197, 202), and the question of when the limit is reached remains unclear. In the consultations on changes in the Sámi Act (Draft Resolution and Bill 2020–2021, p, 28), the emphasis of the Sámi Reindeer Herders Association of Norway (NRL) on legislating central elements of the protection of Article 27 can be viewed as a response to this vagueness. The Ministry's reaction so far is that they will assess the proposal of regulating the material limit of Article 27 in the follow-up of SRUII (Draft Resolution and Bill 2020–2021, p, 104). What this chapter reveals is that the decisions in different cases appear to be less consequent and that the views from case to case differ on past and present negative impacts on reindeer and pastures caused by wind power plants and mitigation measures. Still, while Article 27 does not provide absolute protection, it does contain a "core zone," and interventions in Sámi culture comparable to a denial of cultural practice are constrained by the provision (NOU 2007a, pp. 197, 202, 211; see also Expert group report 2018, p. 45). Or as the Ministry argues in Kalvvatnan

> the individual reindeer herder cannot be denied the right to reindeer herding. Even if a proportionality principle applies within international law, international law sets an absolute limit to what kind of endeavors can be allowed. Where there is reasonable doubt about whether an endeavor can be realized within the boundaries of material protection of international law applying to Indigenous peoples, ordinary societal considerations cannot determine whether a license should be granted or not (my translation).
>
> (OED 2016a, p. 13)

State–right-holders–industry interactions

Structural and processual indicators of general applicability enable the assessment of national legislation and institutionalized policies enacted to address the changed interactions between the Sámi Parliament and state authorities. Sámi politics and the institutionalized relationship between the state and the Sámi Parliament depend heavily on international legal developments (Falch and Selle 2018, pp. 23, 201). In terms of the governance triangle, the interactions between the right-holders as civil society actors and the state as the duty-bearer with specific obligations are especially important, with a focus on substantive human rights. While the general applicability of state compliance in terms of structural and process indicators might convey a story of right-holders' enjoyment of rights, outcome indicators in concrete cases might show that the opposite is true, that the state has failed to comply with human rights. This was not the case in Kalvvatnan, but such a failure can be exacerbated along the civil society–industry axis of the governance triangle, as illustrated by the non-agreement on mitigation measures in the Øyfjellet case. Along this axis, business enterprises might play a role "in both the infringement and fulfilment of human rights" as they have an independent responsibility to respect human rights (Anaya 2013, pp. 14–15). The White Paper (White paper 2019–2020) signals improved Sámi involvement in impact assessment processes:

> License applications which contain documentation of reindeer herding participation in the assessments and an agreement on mitigation and compensating measures will be prioritized in the licensing process over applications which lack sufficient documentation of these concerns (my translation).
>
> (White paper 2019–2020, p. 41)

The Sámi Parliament agrees that both involvement and impact assessments must be improved, but they cannot accept continued wind power development in reindeer herding areas (White paper 2019–2020, pp. 41–42; Ságat, June 22 2020; Sámi Parliament 2020). A core challenge remains: the interaction between right-holders and the industry is often experienced as unequal, whether in terms of costs and capacity to respond to project proposals or challenges associated with language differences, translation and communication (NIM 2019). Indigenous peoples frequently must bargain from a position of disempowerment for their rights to be respected (Wilson 2017, p. 11). The "'right to negotiate' provides only an opportunity to pursue a degree of Indigenous control, it does not guarantee it" (O'Faircheallaigh 2018, p. 4). While the White Paper argues for strengthen Sámi involvement, it has been criticized for lacking commitment to providing the affected reindeer herders with financial and other resources to enable them to follow the processes (Sámi Parliament 2020).

Although there are incentives for companies to improve their performance in protecting and respecting Indigenous rights, the state cannot delegate its human rights responsibility to businesses. States are the primary duty-bearers of human rights obligations under international law. They must take steps to prevent, investigate, punish and redress human rights abuses through legislation and regulation (NIM 2019, pp. 32–33), as pointed out in the *Guiding principles of business and human rights* (2011), adopted by the UN Human Rights Council. Furthermore, if the capacity and resource challenges are not addressed in the relationship between right-holders and companies, the former will continue to find themselves in unequal, uneven and unfair interactions.

Concluding remarks

In the Kalvvatnan case, the argumentation leading up to the Ministry's refusal to license the building of a new wind power plant can be seen as an assessment of precautionary measures taken to avoid violating human rights through government action. This is expressed in the evaluation of the current and past impacts by other industries and infrastructure, the effects of disturbing reindeer herding lands and—as a mitigation measure—the ineffective alternative migration routes. The benefit of the doubt serves reindeer herding due to divergent research results. The government stands by its words by referring to the premises of earlier consultations and acknowledges reindeer herding as an important carrier of South Sámi language and culture. The sum total of the existing interventions and the anticipated disturbances from a new wind power plant risk preventing the maintenance of current levels of reindeer herding in the area.

The review released by the Ministry of Petroleum and Energy draws attention to two central points. Kalvvatnan might serve as a standard of compliance against which other cases can be measured, be it procedural aspects of respecting established principles or how to comprehend the precautionary principle as related to the impacts of the disturbance caused by industry and infrastructure. While Kalvvatnan is an example of the threshold of violating Article 27 and how this Article can delimit the authorities' margin of appreciation, in other cases the government's assessment of Article 27 has led to a different outcome as the reference cases here illuminate. Further research is needed on decision-makers' evaluations in different cases of, for example, the limits of the margin of appreciation, the impact of cumulative impacts and the application of the precautionary principle.

Second, the distinction between structural, process and outcome indicators in a narrative assessment of state compliance is an attempt to contribute to the evaluation of the application of human rights standards in national legislation, state policies and efforts that enable people to participate in processes of implementing human rights and the evaluation of results of compliance with human rights treaties. As already mentioned, while the general applicability of state compliance in terms of structural and process indicators might suggest that the right-holders' rights are respected, outcome indicators of concrete cases may assess state failure to comply with human rights, and reveal relevant features of the results of the duty-bearers' efforts to realize human rights.

Finally, while recognizing that compliance is not an either/or issue but rather an issue of determining to what extent nations comply with their obligations, it is equally important to consider that Indigenous law is developing in interaction between international law and state practice (NOU 2007a, p. 208). There is a discrepancy between formal international legal obligations and actual compliance, but the very effort of commitment "sets processes in train that constrain and shape governments' future behaviour, often for the better" (Simmons 2009, p. 8).

References

Åhrén, M. (2016) *Indigenous peoples' status in the international legal system*. Oxford: Oxford University Press.

Anaya, J. (2013) *Report of the special rapporteur on the rights of indigenous peoples. Extractive industries and indigenous peoples*. A/HRC/24/41. Geneva: Human Rights Council.

Anti, E.R. (2020) *FNs konvensjon om sivile og politiske rettigheter art. 27 som begrensning i ekspropriasjonsadgangen av reinbeiteområder*. [Article 27 of the ICCPR as restriction in the right to expropriate reindeer grazing areas] Master's thesis in Law. UiT Norges arktiske universitet.

Barten, U. (2015) "Article 27 ICCPR: a first point of reference," in H. Caruso and R. Hofmann (eds.) *The United Nations declaration on minorities: an academic account on the occasion of its 20th anniversary (1992–2012)*. Leiden: Brill/Nijhoff, pp. 46–65.

Broderstad, E.G. (2008) *The bridge-building role of political procedures. Indigenous rights and citizenship rights within and across the borders of the nation-state.* Doctoral dissertation. Tromsø: Department of Political Science, Faculty of Social Sciences, University of Tromsø.

Bull, K.S. (2018) "En kommentar til Høyesteretts forståelse av SP artikkel 27 og konsultasjonsplikten," [A commentary to the Supreme Court's understanding of article 27 of ICCPR and the obligation to consult] *Lov og Rett*, 57(8), pp. 504–509.

Consultation Agreement. (2005) *Avtale om prosedyrar for konsultasjoner mellom statlige myndigheter og Sametinget, inngått mellom kommunalt- og regionalministeren og sametingspresident 11. mai 2005.* [Consultation agreement between the Sámi Parliament and Government]

Consultation Agreement. (2009) Konsultasjonsavtale mellom Sametinget og NVE 2009. *Avtale om gjennomføring av konsultasjoner mellom Sametinget og NVE, 31.03.2009.* [Consultation agreement between the Sámi Parliament and the NVE]

Consultation Minutes. (2016) Protokoll. Konsultasjon mellom Olje- og energideparte-mentet og Sametinget om tre vindkraftprosjekter i Nordland. October 25. [Consultation protocol between the Ministry of Petroleum and Energy and the Sámi Parliament on three wind power projects in Nordland].

Consultation Protocol. (2012) Protokoll konsultasjonsmøte med Sametinget. May 5. [Minutes of the consultation meeting between the Sámi Parliament and NVE].

Dalen, E.V. and Villaruel, A. (2019) *Nasjonal ramme for vindkraft. Det folkerettslige vernet av samiske interesser i konsesjonsbehandlingen.* [National framework of wind power. International law protection of Sámi interests in the processing of licenses]. NVE rapport nr 15/2019.

De Beco, G. (2008) "Human rights indicators for assessing state compliance with international human rights," *Nordic Journal of International Law*, 77(1–2), pp. 23–49. Available at: https://doi.org/10.1163/090273508X290681

Draft Resolution and Bill. 2006–2007 Ot.prp.nr.25 (2006–2007) *Om lov om reindrift (reindriftsloven)* [Bill on the reindeer husbandry act] [online]. Available at: https://www.regjeringen.no/no/dokumenter/otprp-nr-25-2006-2007-/id446518/

Draft Resolution and Bill 2020–2021. Prop. 86 L (2020–2021) *Proposisjon til Stortinget, Endringer i sameloven mv. (konsultasjoner)* [Draft resolution and bill, changes in the Sámi Act (consultations)] [online]. Available at: https://www.regjeringen.no/no/dokumenter/prop.-86-l-20202021/id2835131/

Eftestøl, S., Colman, J. and Flydal, K. (2011) *Kalvvatnan Vindkraftverk – KU fagtema reindrift* [Impact assessment, reindeer herding]. Available at: https://webfileservice.nve.no/API/PublishedFiles/Download/200801262/473569 (Accessed June 19, 2020).

Energy Act. (1990) *Lov om produksjon, omforming, overføring, omsetning, fordeling og bruk av energi m.m. (energiloven)* [online]. Available at: https://lovdata.no/dokument/NL/lov/1990-06-29-50

Expert Group Report. 2018, Report on the Minerals Act 2018, appointed by the Ministry of Industry and Fisheries. [online]. Available at: https://www.regjeringen.no/contentassets/e9aea8a3cf974c1995dd1665e1ac0b12/innstillingevalueringminerallovsisteversjon.pdf (Accessed August 22, 2020)

Falch, T. and Selle, P. (2018) *Sametinget. Institusjonaliseringen av en ny samepolitikk.* [The Sámi Parliament. The institutionalization of a new Sámi politics]. Oslo: Gyldendal Akademisk.

Fjellheim, E. (2020) "Through the stories we resist. Decolonial perspectives on South Saami history, indigeneity and rights," in A. Breidlid and R. Krøvel (eds.) *Indigenous knowledges and the sustainable development agenda*. London: Routledge, pp. 207–226.

Fjellheim, S. (1991) *Kulturell kompetanse og områdetilhørighet: metoder, prinsipper og prosesser i samisk kulturminnevernarbeid.* Snåsa: Saemien sijte.

Gulbrandsen, L. H., Indreberg, T. H. J. and Jevnaker, T. (2021) "Is political steering gone with the wind? Administrative power and wind energy licensing practices in Norway," *Energy Research & Social Science*, 74. Available at: https://doi.org/10.1016/j.erss.2021.101963

Human Rights Act. (1999) *Lov 21. mai 1999 nr. 30 om styrking av menneskerettighetens stilling i norsk rett* [online], [Unofficial translation in English: Act relating to the strengthening of the status of human rights in Norwegian law (The Human Rights Act)] [online]. Available at: https://lovdata.no/dokument/NLE/lov/1999-05-21-30

ICCPR. (1966) United Nations Office of the High Commissioner for Human Rights, OHCHR *International Covenant on Civil and Political Rights* [online]. Available at: https:// www.ohchr.org/en/professionalinterest/pages/ccpr.aspx (Accessed January 15, 2020)

ILO. (2016) *Procedures for consultations with indigenous peoples – Experiences from Norway*. Gender, equality and diversity branch. Geneva: International Labour Office [online]. Available at: https://www.ilo.org/wcmsp5/groups/public/---dgreports/---gender/documents/ publication/wcms_534668.pdf (Accessed August, 22 2019)

ILO 169, International Labour Organization Convention. (1989) *C169 – Indigenous and Tribal Peoples Convention (No. 169)* [online]. Available at: https://www.ilo.org/dyn/normlex/en/f ?p=NORMLEXPUB:12100:0::NO::P12100_INSTRUMENT_ID:312314

Indigenous Navigator [online]. Available at: https://indigenousnavigator.org/ (Accessed August 22, 2020).

Jacobson, H. (1998) "Conceptual, methodological and substantive issues entwined in studying compliance," *Michigan Journal of International Law*, 19(2), pp. 569–580.

Jacobson, H. and Weiss, E.B. (1995) "Strengthening compliance with international environmental accords: preliminary observations from a collaborative project," *Global Governance*, 1, pp. 119–148.

Kingsbury, B. (1998) "The concept of compliance as a function of competing conceptions of international law," *Michigan Journal of International Law*, 19(2), pp. 345–372.

Lund, S., Gaup, P. and Somby, P.J. (2020) Vindkraft eller reindrift? [Wind power or reindeer husbandy?] Temarapport 3. Del C. *Erfaringer med vindkraft i reindriftsområder*. Oslo: Motvind Norge & Naturvernforbundet i Ávjovárri.

Ministry of Labour and Inclusion. (2006) Arbeids- og inkluderingsdepartementet. *Veileder for statlige myndigheters konsultasjoner med Sametinget og eventuelle øvrige samiske interesser*. [Guide for state authorities' consultations with the Sámi Parliament and possible other Sámi interests].

Minnerup, G. and Solberg, P. (2011) "Introduction" in G. Minnerupand, P. Solberg (eds.) *First world, first nations. Internal colonialism and Indigenous self-determination in northern Europe and Australia*. Brighton: Sussex Academic Press, pp. 1–21.

NIM. (2016) Norwegian National Human Rights Institution/Norges nasjonale institusjon for menneskerettigheter *Temarapport 2016. Sjøsamenes rett til sjøfiske*. [Thematic report 2016. The Coastal Sámi's right to salt water fishing]. Oslo: NIM.

NIM. (2019) Norwegian National Human Rights Institution and National Contact Point for Responsible Business Norway. *Natural resource development, business and the rights of Indigenous peoples*. Oslo: NIM.

NIM. (2020) Norwegian National Human Rights Institution. *A human-rights based approach to Sámi statistics in Norway*. Oslo: NIM.

NOU. 1984: 18 *Om samenes rettsstilling* [Public report, the first Sámi Rights Committee] [online]. Available at: https://www.regjeringen.no/no/tema/urfolk-og-minoriteter/ samepolitikk/midtspalte/nou-198418-om-samenes-rettsstilling/id622185/

NOU. 2007a. NOU 2007: 13 A *Den nye sameretten: Utredning fra Samerettsutvalget Del I, II og III (kap. 1–15)* [Public report, the second Sámi Rights Committee, part A] [online]. Available at: https://lovdata.no/static/NOU/nou-2007-13a.pdf

NOU. 2007b. NOU 2007: 13 B *Den nye sameretten: Utredning fra Samerettsutvalget Del III* – *kapittel 16–24* [Public report, the second Sámi Rights Committee, part B] [online]. Available at: https://www.regjeringen.no/contentassets/e1e9506bce034637a6cfec8bdf2eec75/no/ sved/nou200720070013000dddpdfs-b.pdf

NRK Sápmi. (2014) *Reindrifta vil seire mot vindkraftutbygging* [online]. [The reindeer herding will defeat the wind power development] Available at: https://www.nrk.no/sapmi/_- reindrifta-vil-seire-1.11853134 (Accessed January, 22 2021)

NRK Sápmi. (2015) *NSR-leder: Uaktuelt med vindmøllepark i Kalvvatnan* [online]. [The NSR chairman: A wind mill park in Kalvvatnan out of question] Available at: https://www.nrk. no/sapmi/nsr-leder_-_-uaktuelt-med-vindmollepark-i-kalvvatnan-1.12404064 (Accessed January 22, 2021)

NVE. (2014) *Norwegian Water Resources and Energy Directorate / Norges vassdrags- og energidirektorat. Bakgrunn for vedtak* [The basis for the decision] [online]. Available at: https://webfi-leservice.nve.no/API/PublishedFiles/Download/200801262/1119588 (Accessed June 19, 2020)

NVE. (2020) Norwegian Water Resources and Energy Directorate. *Øyfjellet vindkraftverk* – *Oversendelse av seks klager på tre vedtak av 18.12.2019 og en klage på vedtak av 08.04.2020, samt begjæring om omgjøring og utsatt iverksettelse, brev til OED av 19.05.2020* [Øyfjellet wind power plant – processing of complaints; application of reversal and postponed implementation] [online]. Available at: https://webfileservice.nve.no/API/PublishedFiles/ Download/201707386/3157966 (Accessed November 25, 2020)

O'Faircheallaigh, C. (2013) "Extractive industries and Indigenous peoples: a changing dynamic?," *Journal of Rural Studies*, 30, pp. 20–30.

O'Faircheallaigh, C. (2018) "Mining, Development and Indigenous Peoples" in S. K. Lodhia (ed.), *Mining and Sustainable Development. Current Issues*, Routledge, pp. 124–142.

OED. (2013) *Ministry of Petroleum and Energy / Olje- og energidepartementet. Vindkraft og kraftledninger på Fosen – klagesak* [Wind power and power lines at Fosen, appeal] [online]. Available at: https://www.regjeringen.no/globalassets/upload/oed/pdf_filer_2/fosen/ vindkraft_og_kraftledninger_pa_fosen_klagesak.pdf (Accessed September 12, 2020)

OED. (2015a) *Ministry of Petroleum and Energy / Olje- og energidepartementet. Hammerfest vindkraftverk – klage på avslag* [Hammerfest wind power plant – appeal of the rejection] March 2, 2015. [online]. Available at: https://webfileservice.nve.no/API/PublishedFiles/ Download/201107331/1383393 (Accessed September 12, 2020)

OED. (2015b) *Ministry of Petroleum and Energy / Olje- og energidepartementet. Aurora Vindkraft AS – klagebehandling av konsesjon til etablering av Fálesrášša vindkraftverk i Kvalsund kommune* [appeal processing of licensed of wind power plant] March 2, 2015. [online]. Available at: https://webfileservice.nve.no/API/PublishedFiles/Download/201002742/1383392 (Accessed September 12, 2020)

OED. (2016a) *Ministry of Petroleum · and Energy / Olje- og energidepartementet. Fred. Olsen Renewables AS – Kalvvatnan vindkraftverk i Bindal og Namskogan kommuner – klagesak* [Appeal case of Kalvvatnan] [online]. Available at: https://www.regjeringen.no/contentassets/2cb 371d9a0204b19a8a914ae830a62ee/vedtak-kalvvatnan.pdf (Accessed September 12, 2020)

OED. (2016b) *Ministry of Petroleum and Energy / Olje- og energidepartementet. Eolus Vind Norge AS. Fred Olsen Renewables AS – Øyfjellet og Mosjøen vindkraftverker – klagesak* [Appeal case of Øyfjellet and Mosjøen] November 16, 2016. [online]. Available at: https://webfileservice. nve.no/API/PublishedFiles/Download/201104174/1910185 (Accessed September 12, 2020)

OHCHR, Office of the United Nations High Commissioner for Human Rights. (2012) *Human rights indicators: a guide to measurement and implementation.* New York and Geneva: United Nations Human Rights Office of the High Commissioner.

Olsen, I.S.M. (2019) *Vindmøller på bekostning av samisk reindrift? Utgjør FNs internasjonale konvensjon om sivile og politiske rettigheter artikkel 27 noen forskjell?* [Wind mills at the expence of Sámi reindeer herding? Does Article 27 of ICCPR make a difference?] Master's thesis in Law. Tromsø: UiT Norges arktiske universitet.

Parliamentary Ombudsman. (2009) *Sivilombudsmannens årsmelding for 2009* SOMB-2009 Dok. 4 (2009–2010) Available at: https://www.sivilombudsmannen.no/wp-content/uploads/2017/04/A%CC%8Arsmelding-for-Sivilombudsmannen-2009.pdf (Accessed September 14, 2020)

Planning and Building Act. (2008) *Lov 27. juni 2008 nr. 71 om planlegging og byggesaksbehandling* [online]. Available at: https://lovdata.no/dokument/NL/lov/2008-06-27-71

Procedures for Consultations. (2005) between State Authorities and The Sami Parliament [Norway] [online]. Available at: https://www.regjeringen.no/en/topics/indigenous-peoples-and-minorities/Sami-people/midtspalte/PROCEDURES-FOR-CONSULTATIONS-BETWEEN-STA/id450743/

Public Administration Act. (1970) *Lov om behandlingsmåten i forvaltningssaker* [online]. Available at: https://lovdata.no/dokument/NLE/lov/1967-02-10

Ravna, Ø. (2014) "The Fulfilment of Norway's International Legal Obligations to the Sámi – Assessed by the Protection of Rights to Lands, Waters and Natural Resources," *International Journal on Minority and Group Rights*, 21(3), pp. 297–329.

Ravna, Ø. (2020) "Sámi law and rights in Norway – with a focus on recent developments" in T. Koivurova, E.G. Broderstad, D. Cambou, D. Dorough, and F. Stammler (eds.) *Routledge handbook of Indigenous peoples in the Arctic.* London: Routledge, pp. 143–158.

Reindeer Husbandry Act. (2007) *Lov 15. juni 2007 nr. 40 om reindrift (reindriftsloven)* [online]. Available at: https://lovdata.no/dokument/NL/lov/2007-06-15-40

Ságat. (2017) *Slik vant de Kalvvatnan* [online]. [The way they won Kalvvatnan] Available at: https://www.sagat.no/index.php?author=87&fra=876 (Accessed October 3, 2017)

Ságat. (2020) *Åpen for mer vindkraft i reinbeiteland* [online]. [Opens up for more wind power on reindeer grazing lands] Available at: https://www.sagat.no/apen-for-mer-vindkraft-i-reinbeiteland/19.22666 (Accessed June 22, 2020)

Sámi Act. (1987) *Lov 12. juni, 1987, nr. 56. Lov om Sametinget og andre samiske rettsforhold* [online]. Available at: https://lovdata.no/dokument/NL/lov/1987-06-12-56

Sámi Parliament. (2014) *Sametinget Brev til Olje- og energidepartementet OED. May 8* [Sámi Parliament's letter to OED]. Karasjok: Sámediggi/Sametinget.

Sámi Parliament. (2020) Sak 14.10.2020, *Sametingets plenum: Vindkraft på land – Sametingets syn.* [Sámi Parliament's plenary, subject number 043/20: Wind power on land – the view of the Sámi Parliament]

Scheinin, M. (2004) "Indigenous peoples' land rights under the International Covenant on Civil and Political Rights," *Aboriginal Policy Research Consortium International (APRCi)*, 195 [online]. Available at: https://ir.lib.uwo.ca/aprci/195

Semb, A.J. (2001) "How norms affect policy: the case of Saami politics in Norway," *International Journal on Minority and Group Rights*, 8(2–3), pp. 177–222.

Simmons, B.A. (2009) *Mobilizing for human rights. International law in domestic politics.* Cambridge and New York: Cambridge University Press.

Skarin, A., Sandström, P., Alam, M., Buhot, Y. and Nelleman, C. (2016) *Renar och vindkraft II – Vindkraft i drift och effekter på renar och renskötsel.* [Reindeers and wind power II – Wind power in operation and effects on reindeers and reindeer herding] Rapport 294. Uppsala: Sveriges lantbruksuniversitet SLU (Swedish University of Agricultural Sciences).

Smith, C. (2014) "Fisheries in coastal Sami areas: Geopolitical concerns?," *Arctic Review on Law and Politics*, 5(1), pp. 4–10.

Swepston, L. (2020) "Progress through supervision of Convention No. 169," *The International Journal of Human Rights*, 24(2–3), pp. 112–126

Ulfstein, G. (2013) *Samiske folkerettslige rettigheter ved naturinngrep. Utredning for Olje- og energidepartementet i forbindelse med utbygging av kraftledninger og vindkraft.* [International legal rights of the Sámi. A report to the OED in connection with development of power lines and wind power] Oslo: Olje- og energidepartementet.

UN Guiding principles of business and human rights. (2011) [online]. Available at: https:// www.ohchr.org/documents/publications/guidingprinciplesbusinesshr_en.pdf (Accessed November 19, 2020)

UN Human Rights. n.d., *Office of the High Commissioner* [online]. Available at: https://www. ohchr.org/EN/HRBodies/CCPR/Pages/CCPRIntro.aspx (Accessed July 19, 2020)

White paper. 2019–2020, Meld. St. 28 (2019–2020) Meldinger til Stortinget, *Vindkraft på land. Endringer i konsesjonsbehandlingen* [Wind power on land. Changes in the processing of licensees] [online]. Available at: https://www.regjeringen.no/no/dokumenter/meld.-st.-28-20192020/id2714775/

Wilson, E. (2017) *What is social impact assessment?* Report of project on Indigenous peoples and resource extraction in the Arctic: evaluating ethical guidelines. Ájluokta/Drag: Árran Lule Sami Centre.

3 Reindeer husbandry vs. wind energy

Analysis of the Pauträsk and Norrbäck court decisions in Sweden

Dorothée Cambou, Per Sandström, Anna Skarin and Emma Borg

Introduction

The rights of the Indigenous Sámi people to conduct reindeer husbandry is based on immemorial prescription and customary law (SFS 1971:437; SOU 2020:73). It is enshrined in the Swedish framework for land use planning and decision-making, which also clearly stipulates that all owners and stakeholders of the land within the Reindeer Husbandry Area must safeguard and respect the grazing rights as defined in the Swedish Constitution (SFS 1974:152, 2 kap. 17 §), the Reindeer Husbandry Act (SFS 1971:437) and the Swedish Environmental Code (SFS 1998:808). Furthermore, the right of the Indigenous Sámi people to conduct reindeer husbandry is also a human right that is recognized at international and national levels, as part of the rights of Indigenous peoples (Ivan Kitok *vs.* Sweden, 1988; Prop. 1976/77:80). Yet, Sámi reindeer husbandry, which has a long history of colonial subjugation, forced adjustments and adaptations, continues to be challenged by ever increasing co-occurring land uses (Lantto and Mörkenstam, 2008; Lawrence, 2014; Össbo and Lantto, 2011; Sköld, 2015; Sandström, 2015). Starting with settlers and tax collectors in the eighteenth century and followed by epochs of forestry, hydropower and mining intrusions (Sandström et al., 2016; Skarin and Åhman, 2014; Vistnes and Nellemann, 2008; Klein, 1971), wind power development is now the latest activity affecting the conduct of reindeer husbandry on traditional Sámi lands.

The negative impact of wind energy projects on reindeer husbandry has been addressed and recognized in several recent studies (Skarin et al., 2015; Skarin, Sandström and Alam, 2018; Skarin and Alam, 2017). Wind energy constitutes a current form of industrial development that jeopardizes reindeer husbandry as a traditional livelihood of the Sámi and their rights as an Indigenous people (Anaya, 2011; Tauli-Corpuz, 2014). In this regard, the development of wind energy is considered by Sámi representatives as a form of neo-colonialism and has triggered an increasing number of lawsuits in both Sweden and Norway, leaving the judicial system to tackle and resolve the issue (Cambou, 2020). In this chapter, we examine this topic through the analysis of two court decisions issued in the Norrbäck and Pauträsk cases, which concern the establishment of two wind energy projects within an important area of reindeer husbandry on the winter pastures of Vapsten reindeer herding community in northern Sweden (Figure 3.1).

DOI: 10.4324/9781003131274-3

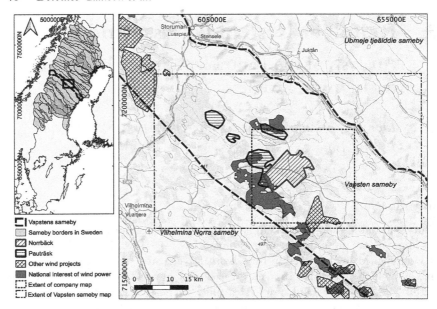

Figure 3.1 The projects areas of Norrbäck and Pauträsk. ©Lantmäteriet.

Several reasons have motivated the selection of these cases. First, the Norrbäck and Pauträsk cases epitomize the decisions of most court cases concerning legal disputes between reindeer herding communities (*samebyar* in Swedish plural and hereafter named "*sameby*," in singular) and wind energy developers. As in the majority of court decisions, these cases, too, resulted in the authorization of the wind energy projects despite impacts on reindeer husbandry (Cambou, 2020). Second, both cases illustrate the complexity of assessing the impact of wind energy projects on reindeer husbandry. Moreover, the specificity of the cases is also linked to the fact that the land and environmental court (*in Swedish* Mark och miljödomstolen) and the court of appeal (*in Swedish* Mark och miljööverdomstolen) did not agree in their final decisions. This raises questions about the courts' assessment and decision to license the projects. Finally, we selected these cases because two of the authors of this chapter participated in the case proceedings as expert witnesses, providing important insights into the background and final decisions of the court cases.

Drawing on the expertise of each author in the field of law, ecology, reindeer husbandry and land use planning, the goal of this chapter is to examine the content of the court decisions and analyze the implications for the protection of the Sámi right to practice reindeer husbandry. We have used working translations of the courts' decisions, academic and legal documents relevant for the interpretation of the decisions and informal facts provided by key informants such as the lawyers of the Vapsten *sameby*. The analysis thus focuses on how information from the companies' environmental impact assessments (EIA) and testimonies by reindeer herders and scientific expert witnesses was interpreted and used en route to the final rulings by the courts.

The first section of the study provides a contextual background, including the legal framework applying to the issue. Section Two examines the courts' assessments and presents an analysis of the decisions of both the land and environmental court and the appeal court, and the underlying factors behind the diverging assessments. The final section discusses the limitations of the appeal court decisions to protect the right of the Sámi to practice reindeer husbandry. Here, we analyze the assessment of the court and its application of the principle of sustainable development with regard to the human rights of the Sámi as an Indigenous people. In line with the theoretical underpinnings of this tome, the chapter also considers the role of the judiciary system and the principle of sustainable development as second and third orders of the meta-governance system of conflict mediation between the promotion of renewable energy and the protection of the Sámi Indigenous people's right to practice reindeer husbandry.

Cases background

In 2015, Vattenfall Vindkraft Sverige AB and Hemberg Energi AB submitted their respective applications to the County Administrative Board (CAB) of Västerbotten for permits to construct two adjacent wind energy projects with up to 120 wind turbines in the Pauträsk area and 55 turbines in the Norrbäck area. In accordance with their "monopoly planning right," the municipalities of Lycksele, Storuman and Vilhelmina had already approved the permit applications, while the regional County Administrative Board initially rejected both applications due to the projects' impact on conflicting interests, which included nature conservation, the cultural environment and reindeer husbandry. This situation therefore opposed regional and municipal authorities. It also opposed national and local interests. While small portions of the project areas have been designated as being of national interest for wind energy, the two areas have always been used as winter pasture by Vapsten *sameby*, and is also close to the winter pastures of Vilhelmina Norra *sameby*. Furthermore, the project areas are located close to several movement and migration routes designated as national interest for reindeer husbandry (Figure 3.1 and 3.2). The Vapsten and Vilhelmina Norra *samebyar* are among the 51 *samebyar* covering the northern half of Sweden, and they both have an exclusive right to carry out reindeer husbandry on their lands. The traditional territory of the Vapsten *sameby* covers around 10,000 square kilometers, of which 7,000 square kilometers are winter pastures. Several industrial projects have already been approved by the State on Vapsten *sameby*'s traditional territory, including other wind projects. Consequently, and as noted by the United Nations Committee on the Elimination of Racial Discrimination (CERD, 2020, p. 2),

> a large part of this territory has already been taken from the reindeer herding community and its pasture land is constantly decreasing, which is creating a real threat to reindeer herding and placing enormous psychological pressure on the community's members.

During the application process, Vapsten *sameby* was scarcely consulted by the companies and duly opposed the application for the establishment of the two wind

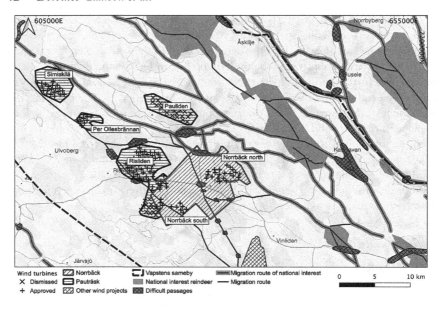

Figure 3.2 The overlapping interest areas. ©Lantmäteriet.

projects (Omma, 2021). That the company did not seek to obtain free, prior and informed consent from Vapsten *sameby* in accordance with human rights standards did not stop the application process. In Sweden, the legislation for licensing wind energy and other industrial activities recognizes the right of the public to be consulted, but it lacks emphasis on the rights of the Sámi people to participate in decision-making processes (Allard, 2018; Larsen and Raitio, 2019, p. 15). More precisely, the legislation does not single out the right of the Sámi to be consulted as an Indigenous people and "leaves a wide margin of appreciation to the developers about the way to organize and implement consultations" (Cambou, 2020, p. 320). Several scholars and human rights institutions have criticized the Swedish legal framework for failure to guarantee the rights of the Sámi people to free, prior and informed consent, which ensures the right of Indigenous peoples to have a real say in developments affecting their traditional lands (Allard, 2018; Larsen and Raitio, 2019; CERD, 2018). The lack of protection of the rights of the Sámi, including the lack of recognition of their traditional right to land and resources, is one of the factors explaining an increase in the number of lawsuits between *samebyar* and wind energy developers. In the absence of recognition of Sámi territorial rights and adequate consultation, the court has become the ultimate governance arena where conflicts between opposing parties and interests can be mediated and decided.

In the Norrbäck and Pauträsk cases, the main issue pertained to the impact of the wind projects in an important area of reindeer husbandry as opposed to the value of the area for establishing wind energy projects. In their testimony, the Sámi reindeer herders from Vapsten *sameby* argued that the impact of the project would be too far-reaching in an area that was considered central as winter pasture. At present, this region connects several grazing areas and allows the free roaming of the reindeer. The herders also explained that viable migration routes through the area were necessary to enable reindeer migration between the winter and summer pastures. The area also provides good grazing conditions in winters with otherwise challenging grazing conditions. The rapid climate change, especially pronounced in the Arctic and subarctic regions, often causes difficult grazing conditions during winters due to rain-on-snow events, which create ice-crust layers in the snow as normal freezing temperatures return, making forage inaccessible to reindeer (Forbes et al., 2016). During such winters, upland and rugged terrain usually provide better grazing conditions thanks to a more stable temperature and increased topographical variation. Such conditions were present in both the proposed wind energy development areas. In the court cases, the reindeer herders contended that sustainable reindeer husbandry was based not only on grazing resources within designated national interest areas, but each *sameby* also depended on areas in well-connected landscapes where it was possible to find good grazing conditions in varied weathers, as in both Norrbäck and Pauträsk.

In contrast, the companies disputed both the importance of the area for reindeer husbandry and the level of impact caused by the development projects. Presenting their arguments individually, the companies argued that their project areas did not constitute important grazing pastures and that the reindeer herders had not demonstrated that continuous migration on foot had occurred in the area during the last twenty-five years. Instead, the companies stated that reindeer herders often had to use trucks to ensure the transportation of reindeer through the area. Both companies also argued that the operation of the movement and migration routes would not be affected and that their assessment showed that the level of project impact was not significant. However, it is not clear from the court documents from where the companies' gained expert knowledge about how reindeer husbandry is carried out in the area. Against the companies' positions, the reindeer herders argued that the area was indeed important for grazing and that winter migration by foot and stationary grazing still occurred in the area, which should therefore be preserved and protected for the future. The Sámi Parliament supported the arguments of the *sameby* in their testimony to the court, emphasizing the importance of the area in securing the maintenance of reindeer husbandry especially in times of climate change. Finally, the neighboring Vilhelmina Norra *sameby* argued that the projects risked affecting an area of importance to them. In particular, the planned wind energy projects would make reindeer from Vapsten *sameby* deviate into their adjacent grazing areas.

Together, the conflicting positions between the *samebyar* and the companies reflect some of the background against which the courts had to arrive at a decision. To reach a decision, the court had to weigh *expert opinions* about the needs of reindeer husbandry and the impacts of wind power developments as presented by the *samebyar* vs. the companies'. It is important to note that the companies' EIA addressed their projects as completely isolated cases with separate impacts on reindeer husbandry, even though the projects were located only five kilometers apart. This scale of assessment differs sharply from the *sameby*'s description of impacts that did not separate the impacts between the two projects.

Figures 3.1 and 3.2 illustrate the wind development projects in relation to reindeer herding and provide a visual background to understanding the issue as it has been addressed by the courts. The mappings of the reindeer grazing areas come from the publicly available iRenmark database and presented together with data from RenGIS (Sandström 2015). All mappings describing the land uses of reindeer husbandry were carried out independent long before the wind development plans became common knowledge.

Legal background

The legal background to the Norrbäck and Pauträsk court rulings is based mainly on the environmental code adopted by Sweden in 1999 (SFS 1998:808) and the case law that has since interpreted the stipulations of the code. The Environmental Code sets several directing principles, among them the promotion of sustainable development, as a basis to regulate the launching of industrial activities such as wind energy projects (Chapter 1, Section 1). Although the purpose to promote wind energy is not explicitly stated, the code gives "[p]reference to renewable energy sources." The court decisions have stressed several times that increased energy production based on wind power can contribute to achieving the environmental code's goals for sustainable development. Accordingly, the court argues that "it is therefore essential that areas suitable for wind power production can be used for this purpose," but it also reiterates that the promotion of wind energy must "take place with due regard to the protection interests of the site" (Norrbäck and Pauträsk cases 2019). In other words, the promotion of wind energy in Sweden is encouraged and regulated by law. In accordance with the Environmental Code, wind projects must comply with certain environmental requirements, including basic and specific resource management provisions and localization requirements as well as the protection of certain interests such as those of reindeer husbandry.

While the legal framework regulating the wind energy process has been described elsewhere (Pettersson et al., 2010; Larsson and Emmelin, 2016; Solbär, Marcianó and Pettersson, 2019), it pays to recall the rules concerning the protection of reindeer husbandry, which play an instrumental role in the decisions to authorize activities on the traditional lands of Sámi reindeer herders. Section 5 of Chapter 3 of the Environmental Code stipulates: "Land and water areas that are important for reindeer husbandry, ...*to the extent possible*, shall be protected against measures that may *significantly* interfere with the operation of these industries" (emphasis added). This provision seeks to protect generally the right of Sámi reindeer herders to use their traditional land and is considered to meet the commitments Sweden has under Article 27 of the International Covenant on Civil and Political Rights (ICCPR) in relation to the right of the Sámi people to culture. Paragraph 2 of Section 5 in Chapter 3 of the code further mentions "[a]reas that are of national interest for the purposes of reindeer husbandry." This provision implies that areas that are designated as being of national interest for reindeer husbandry will be protected against activities that may *significantly* interfere with the interest of the reindeer industry, whereas other areas are usually safeguarded only "to the extent possible." However, when an area is concomitantly designated as being of national interest for both wind energy projects and reindeer husbandry, the code indicates that priority shall be given to the purposes that are most likely to promote the sustainable management of land (Chapter 3, Section 10). This vague formulation stands in the way of any predictions of how governing institutions, including the courts, will balance competing interests between different land uses that equally promote sustainable development (Pettersson, 2008, p. 40).

Nonetheless, the application of these provisions in the legal disputes concerning wind energy development and its impact on reindeer husbandry has started to provide some insights into the ways the court interprets the provisions in practice. For example, in a growing number of lawsuits (at least twenty-nine since 2008) the Land and Environmental Court of Appeal has begun to assess what impact the establishment of wind power could have on conducting reindeer husbandry. In its latest rulings, the court has confirmed that research clearly indicates that wind energy projects lead to avoidance effects of reindeer at significant distances from wind power (Pauträsk case 2019), suggesting that wind energy projects can have negative effects on reindeer husbandry. However, according to several court rulings, it is difficult to specify the exact distance when the effects of wind energy projects will be such that the impact on reindeer herding is clearly indisputable. As a result of this uncertainty, most controversies in many legal disputes revolve around two central issues. The first is that of knowing whether the impact of wind energy projects can significantly affect and jeopardize reindeer husbandry. The second is to decide which interests best promote the objective of sustainable development when wind energy projects conflict with the needs of reindeer husbandry.

Although the case law concerning these issues is still developing, a survey of the decisions since 2008 demonstrates that the court has rarely rejected permits because of the impacts that wind projects have on reindeer husbandry (Cambou, 2020). Among these decisions is the Ava Case, which concerns a project application for nineteen wind turbines in the area of Vilhelmina Norra *sameby*. In its decision the court annulled the construction permit after concluding that the impacts of the project on reindeer husbandry would be significant since the Gabrielsberget wind project operated in the same area (Ava Case, 2018). Given the great risk of putting an end to traditional reindeer husbandry in the area if the wind energy project were expanded according to the application, the court considered that the interest of protecting the reindeer husbandry industry outweighed that of wind power expansion. The assessment of the court was based on reindeer herders' testimonies and scientific studies describing the impacts of the existing turbines on reindeer in the Gabrielsberget area. These studies provided evidence that the existing wind project made it significantly more difficult to conduct reindeer husbandry in the area and entailed the risk of long-term deterioration of the viability of grazing lands. The court also concluded that no safeguarding measures would be able to counteract or prevent the inconvenience for reindeer husbandry if the area's natural function disappeared. Consequently, the court rejected the permit application for the new project.

Apart from the Ava Case, however, the Swedish courts have seldom assessed that a wind energy project's impacts on reindeer husbandry would be sufficiently significant to justify the rejection of a permit. In effect "it appears that the courts are more likely to reject a permit for wind energy if compelling evidence demonstrates that reindeer husbandry might completely cease as a result of the disturbance caused by the project in question" (Cambou, 2020). However, because it is difficult to prove in practice that a wind project would cause the cessation of reindeer husbandry, the court has in most cases decided to grant a license provided that measures will be taken to mitigate the potential impacts (Cambou, 2020). In other words, the court often concludes that the impact of wind energy projects on the conduct of reindeer husbandry can be mitigated and compensated by measures which will ensure that reindeer husbandry does not cease in practice. The Pauträsk and Norrbäck decisions thus illustrate cases in which the court has authorized wind energy projects despite the impacts on reindeer husbandry, but the two cases also raise questions about the assessment of the court and the justifications provided to allow a permit in view of the impacts on reindeer husbandry. These questions will be the focus of the following sections.

Court assessments

In the adjacent Norrbäck and Pauträsk cases, the environmental and appeal courts respectively issued two separate decisions in 2017 and 2019 in response to the permit applications of the two wind energy companies. Because both cases involved identical plaintiffs and similar application areas, the courts replicated their arguments in both cases with almost identical wordings. And yet, while the decisions in the Norrbäck and Pauträsk cases were based on similar grounds, the land and environmental court came to a different conclusion than the court of appeal. The differences are reviewed in the next sections. First, the analysis focuses on the judgments of the land and environmental court, which rejected the permits for wind projects. The second section then examines the decisions of the court of appeal and its justification to authorize the wind projects despite the impact on the practice of reindeer husbandry and other conflicting interests.

Decisions of the land and environmental court (2017)

In both the Norrbäck and Pauträsk cases, the court agreed that the area targeted by the wind energy developers constituted a critical area for several opposing land uses, including reindeer husbandry. The potential significance of the impacts of the projects on reindeer husbandry was in fact one of the central assessments for the court. In accordance with the Environmental Code, the court was specifically bound to assess whether any *significant* impact could threaten reindeer husbandry in the area. Indeed, although the area was of importance for reindeer husbandry, only areas affected by *significant interference* come under the protection of the Environmental Code (Chapter 3, Section 5). The court pointed out that the number of studies on the impact of wind power developments on reindeer has gradually increased and that the studies show avoidance effects, albeit with different disturbance zones (Skarin et al., 2015; Skarin and Alam, 2017; Skarin et al., 2016; Skarin, Sandström and Alam, 2018). As stated by Skarin in her expert opinion in the Norrbäck and Pauträsk cases, studies show that reindeer avoid or decrease their use of wind energy sites, both within calving and winter pastures, and a decrease in use has been observed up to three to five kilometers away from wind energy sites. Skarin's testimony also acknowledges a likely variation in reindeer avoidance of wind farms depending on local conditions. According to studies, then, wind energy projects have a negative impact on reindeer habitat selection and behavior.

However, the court also remarked that there was uncertainty about the extent of such impact. These uncertainties were also evidenced throughout the arguments of the companies and the *sameby*, supporting different views about the distance at which wind energy projects could cause negative effects on reindeer husbandry. Whereas the evidence presented by the *sameby* supported the view that avoidance effects could be seen at a distance of three to four kilometers from wind energy projects, the company referred to recent studies which showed that negative effects could only occur within a one-kilometer zone. In the light of these uncertainties but given the importance of the area for reindeer husbandry, the court found—on the available research basis—that it could not rule out the possibility of significant harms to reindeer husbandry in Vapsten *sameby* caused by the establishment of wind power within the license application area (Pauträsk case 2017).

Subsequently, another point for the court to assess was whether any protective measures could safeguard the continuation of reindeer husbandry against the impacts of a wind energy project. The court noted that the companies had expressed a positive attitude to discussing mitigation measures. However, the court also established that the companies had not been able to present safeguards and precautionary measures that would ensure that the wind energy projects could coexist with reindeer husbandry. According to the court, "one of the basic prerequisites for reindeer husbandry as a general interest is that the industry is practiced in the traditional Sámi way, where the reindeer graze natural pastures on large continuous lands." (Norrbäck Case.) The court indicated that financial compensation to an individual *sameby* to replace the loss of natural pastures and substitute natural nourishment with feed on a continuous basis, as suggested by the companies, could not be such a measure. As a consequence, the court decided that the measures provided by the companies could not ensure that the *sameby* could continue to carry out traditional reindeer husbandry based on natural pastures and maintain its economic viability in accordance with their traditional values.

The land and environmental court therefore decided in both the Norrbäck and Pauträsk cases that the interests of energy production should be disregarded to protect the opposing interests, namely the protection of nature, cultural environment and reindeer husbandry. Together these interests would better promote the long-term management of the area compared with the wind energy projects. On July 3, 2017, the Umeå Land and Environmental Court rejected both permit applications for the establishment of wind energy projects in the Norrbäck and Pauträsk areas.

Decisions of the Land and Environmental Court of Appeal (2019)

After an appeal was launched on September 4, 2019, the court of appeal overturned the decisions of the land and environmental court and judged that it would be permissible for the companies to construct and operate their wind energy projects. The appeal court made its decisions in light of the companies' adjustments to the applications in terms of the projects' scope and localization of the wind turbines. The main purpose of some of these adjustments was to accommodate the protection of the natural and cultural values of the environment in the project area, while other adjustments were directly linked to the protection of reindeer husbandry. In contrast, the *sameby* contested the company investigations, which according to the Sámi did not give a correct picture of reindeer land use in the area.

As a primary adjustment to accommodate the interests of reindeer husbandry in the Pauträsk case, the company suggested removing the Pauliden sub-area from the project application. By preventing the construction of any wind turbine in this sub-area, the company argued that the project would be adequately distanced from areas that are regularly used by reindeer. Exempting the Pauliden area from the project application, the company claimed, would ensure that all remaining sub-areas would be located in the southern part of Pauträsket. This would guarantee that the reindeer's main movement and migration routes of national interest would not be flanked with wind turbines. The company also pointed out that this adjustment would increase the distance between the project sites and the nearest national interest main migration routes from 800 meters to 1,600 meters. According to the company, these adjustments were specifically made to meet the demands of the *sameby* and ensure that the project would not affect their activities. The County Administrative Board had initially rejected the permits, but now decided to support such an adjustment and the licensing application in the appeal case. In agreement with the company, the County Administrative Board indicated that the exclusion of Pauliden from the project application constituted "a better alternative with less impact on opposing interests." Provided that adequate conditions were established to protect reindeer husbandry, the County Administrative Board also argued that the proposal to exclude Pauliden would enable both reindeer husbandry and the development of renewable energy on the site (Pauträsk case 2019), and therefore found in favor of permit authorization.

In its decision, the court concluded that the exemption of the Pauliden sub-area was a valid means of ensuring that the Pauträsk project would not significantly impact reindeer husbandry. Despite the lack of consensus about the importance of the Pauträsk area for reindeer husbandry, the court agreed that the Pauliden sub-area was of importance for reindeer husbandry and therefore required specific protection. For the court, the importance of this sub-area—800 meters away from the original planning site—was linked to the fact that the construction of wind turbines in Pauliden would risk obstructing reindeer husbandry to a significant degree, mainly through disruptions to the movement and migration routes classified as being of national interest. Also, the hearing with the scientific expert Anna Skarin and supporting research had, according to the court, demonstrated the existence of a risk that the reindeer could be negatively affected if wind turbines were to be constructed in Pauliden. These factors justified the exclusion of the area from the project application.

In contrast to the *sameby*'s testimony, the court considered that the other sub-areas of the Pauträsk project were not of importance for reindeer husbandry. The decision was based on the court's own interpretation of GPS data and their interpretation that the reindeer herders' testimony did not provide any evidence of continuous use of the Risliden, Per-Ollesbrännan and Simiskilä areas during the last twenty-five years.On this basis the court found court thus found that the establishment of the wind energy projects would not significantly obstruct the practice of reindeer husbandry in these areas. Although the court agreed that the project might cause some difficulties for the *sameby* to conduct reindeer husbandry, it also argued that any disruption in the use of winter pastures and migration routes would be considered acceptable. In its decision, it suggested that two factors mitigated the disruption: the importance of increased production of renewable energy to reach nationally set renewable energy targets, and the fact that the company had undertaken to compensate the *sameby* for the difficulties that the project might incur. Apart from Pauliden, the court therefore decided that there were no obstacles to granting a permit for the construction of wind energy projects in the Pauträsk area.

In the Norrbäck case, the company similarly suggested decreasing the scope of the project and designated several exemption areas for construction to mitigate the impact of the project on natural and cultural environmental values and on reindeer husbandry. Specifically, the companies proposed the removal of the Björnberget site in their revised application to take into account the protection of reindeer husbandry. This adjustment was also considered adequate by the court, as the project would then be located at "an appropriate distance" from the main movement and migration routes of national interest. As in the Pauträsk application, the court considered that the rest of the project area did not compromise reindeer husbandry, because the company had provided adequate measures to ensure that the project would not significantly impact the activity in the other areas. These measures included an adaptation of the construction work schedule during two grazing seasons when the area was not used for reindeer husbandry. It also included some financial support to compensate for the loss of pasture, the cost of additional labor for the reindeer herders, the purchase of a snowmobile and the construction of a corral. Given the importance of increased production of renewable energy and the fact that the company had also undertaken to compensate the *sameby*, the court concluded that the establishment of the wind energy project in the Norrbäck area did not create any obstacles to reindeer husbandry. Similarly to the Pauträsk case, the court of appeal therefore decided to grant the permit for the establishment of the wind energy project in the Norrbäck area.

Limits of court decisions

While the court decisions provide a basis to conciliating the interests of wind energy development with the protection of reindeer husbandry, the capacity of the judiciary to ensure the protection and sustainability of reindeer husbandry has been questioned (Cambou, 2020). Drawing on the Norrbäck and Pauträsk decisions, we will discuss some of the limitations of the court decisions which address the conflicts between wind energy developers and *samebyar*. In particular, the following section focuses on the role of the court and its assessment of the impacts of the wind energy projects on reindeer husbandry, and specifically the technical uncertainties that the decisions raised. Beyond the technicality of the assessment, the analysis also considers the application of the principle of sustainable development as a meta-governance tenet and its role in guiding the court decisions. While the court decisions support the objective of promoting sustainable development in accordance with the Environmental Code, this section questions whether the court decisions to promote sustainable development at the national level undermine the sustainability of reindeer husbandry based on the Sámi rights as an Indigenous people at the local level.

Uncertainties of court assessment

From the outset, it is important to note that the decision of the environmental court of appeal in the Norrbäck case was not unanimous. In a dissenting opinion, Chief Justice Roger Wikström, one of the three judges appointed to the case, opposed the final decision issued by the court of appeal. Wikström raised two main arguments against the ruling: first, he found that a part of the area in Norrbäck, which could "be described as a backbone for the biodiversity in the county" was not adequately considered in the decision of the court and should have "in this case be[en] given priority over the production of renewable energy." Second, and more importantly for this analysis, Wikström also criticized how the court of appeal had struck a balance between the interests of renewable energy and the protection of reindeer husbandry. He specifically underlined that "areas that are important for reindeer husbandry are to be protected as far as possible against measures that can significantly hamper the business activity." For this purpose, he also explained that protecting nature in the area against wind energy projects the size of the Norrbäck project would also have safeguarded continued reindeer husbandry in the area. In his view it was clear from the information received "that the area is important for the reindeer industry, especially for Vapsten *sameby*, particularly as a number of other developments have been allowed in their area." Considering the negative impact that the establishment of the wind energy project would entail for the natural conditions of the area and for the reindeer industry, Judge Wikström therefore concluded that the application for the establishment of the Norrbäck project and the appeal should have been rejected in its entirety.

This dissenting opinion raises several uncertainties about the assessment of the court, including its methods and scale of appraisal. First, the dissenting opinion challenges the decision of the court to disregard the importance of the entire area for the conduct of reindeer husbandry. The court's reliance on the description of certain areas and migration routes as being "of national interest" to justify the protection of certain sub-areas but exclude others is questionable. In line with Wikström's dissenting opinion, reindeer herders have repeatedly argued that all grazing lands are of great importance during certain periods. In contrast, the court assessment supports an ecological understanding that fragments land use and decreases landscape connectivity by authorizing new projects and forces reindeer herders to adapt to new circumstances. Land fragmentation and the failure to provide for quality grazing areas have both been underlined as a threat to the Sámi way of life (Löfmarck and Lidskog, 2019).

Second, the court assessment of the impact of the project on reindeer husbandry and its conclusion that the project would not significantly affect reindeer husbandry is also questionable. In fact, it is striking how little information has been provided in the decisions about the ways the court of appeal has carried out its assessment of the impact of the wind energy projects. In particular, it is not transparent what scientific research or other evidence were taken into account in either of the two cases to motivate the assessment that the impact of the project outside the exemption locations would not risk obstructing reindeer herding to a significant degree in the exploitation area. The final court rulings also lack details to explain how the new project application and its adjustments provided better protection than the former application. In this regard, whether the companies' mitigation measures meet the threshold of appropriate precautionary efforts to avert the negative effects of wind power development on reindeer husbandry can certainly be challenged.

Looking at the assessment methods, it also appears that the court decisions have made limited and inadequate use of maps to improve their understanding of the cases. The maps presented by the companies to the courts were produced on a local scale, making evaluation difficult on the landscape scale (Figure 3.1). This becomes clear from the reproduction of the map provided by the companies in the Norrbäck case. By submitting a map focused solely on the wind energy project areas, the companies limit the opportunity for both the County Administrative Board and the courts to fully understand and examine the aggregated impacts of all developments in the area. In opposition, the reindeer herders argue that the maps must show larger areas to give a full overview that illustrates the reindeer's use of the landscape. The maps supplied in court by the *sameby*, which describe the entire winter grazing area and show both adjacent wind energy projects (Figure 3.1) stand in sharp contrast to the local maps presented by the companies. In addition, the limited use of maps and the emphasis on maps at a local scale together with a reliance on verbal testimonies steered the court assessment toward an incomplete description of the impacts of the projects. This makes it more difficult to explain complex and conflicting land uses in the area (Sandström, 2015) and the importance of landscape and ecological connectivity. These limitations also become apparent when two projects such as Norrbäck and Pauträsk are evaluated separately and independently of each other, even though they are located only about five kilometers apart and consequently have overlapping impacts. Furthermore, several other wind energy projects approved within the Vapsten *sameby* winter grazing area (or with pending applications), were outside the scope of the maps presented by the companies (Figures 3.1 and 3.2). The scale of the assessment therefore raises important issues, especially as it fails to uncover the totality of the impacts that cumulatively burden reindeer husbandry in the area (Larsen et al., 2016).

In fact, the appeal court's decisions largely left undiscussed the issue of cumulative impacts. Vapsten *sameby* is 33 kilometers wide at the latitude of the projects. According to the development plan, the wind energy projects would occupy nineteen kilometers of this width, leaving a fourteen-kilometer corridor outside the wind energy projects (Figure 3.2). Within this fourteen-kilometer wide corridor, there is ongoing industrial forestry, the E12 highway, the Lycksele to Storuman railroad and several hydroelectric developments along the river Umeälven. Before the wind energy project applications, these other land uses already prevented or hampered the use of several migration and movement routes of national interest. These prior land uses had already led to major impacts, making co-existence with additional new wind power developments particularly difficult to achieve. Such complex land use situations call for implementation of the precautionary principle when assessing additional new development projects. Yet, the court paid little or no attention to the cumulative impacts of the new wind energy projects in relation to existing development projects.

All in all, the court decisions do not entail a comprehensive examination of the sustainability of reindeer husbandry on the project sites and remaining lands as a whole. As a result, there is significant uncertainty about the impact of the wind projects on reindeer husbandry. In this context, the argument that functioning and sustainable co-existence between reindeer husbandry and wind turbines is possible in Sweden can be called into question, especially when the cumulative impacts of all ongoing projects are considered. As explained by the Sámi Parliament in the Norrbäck case, the area of available grazing pastures is shrinking to such an extent that it threatens the economy of reindeer husbandry and exacerbates its ecological vulnerability. In the absence of scientific certainties regarding the possible impact of the wind projects on reindeer husbandry, the court decisions therefore pose a risk for the future conduct of reindeer herding in the area. This is even more the case as there is a lack of evidence that the companies' mitigation measures will adequately alleviate damages in practice.

Balancing of national and local sustainability interests

Whereas the precautionary principle did not guide the court in its decision-making, sustainable development is the meta-governance principle that has informed the court in its decision regarding which public interest should prevail in the management of the land. Although sustainable development does not have a precise definition, it is clear from the court statements that the national policy goal to promote renewable energy, including wind energy, is in accordance with the objective to promote sustainable development in Sweden. The importance afforded to wind energy for promoting sustainable development and the meeting of environmental targets set by parliament in 2016/17 recurred in the court decisions. In contrast, it is unclear from the cases how much the practice of reindeer husbandry as a sustainable activity is valued by the court in comparison to wind energy. Despite the disruption wind energy projects may cause to the environment and reindeer husbandry, the court argued that these risks "may be considered acceptable given the weight of increased production of renewable energy" (Norrbäck case 2019). In this context, the court of appeal asserted the importance of achieving the national renewable energy goals as an underlying motivation for the decisions in favor of wind energy developments.

The decision to promote wind energy to the detriment of reindeer husbandry may be valid from the standpoint of national law and Sweden's ambitious goal to achieving negative carbon emissions by 2045, but it can nonetheless be more fundamentally challenged from the perspective of the consequences for local sustainability and the right of the Sámi to conduct reindeer husbandry. The court decisions emphasized an understanding of sustainable development which tends to promote sustainable development at a national level rather than ecological and cultural preservation at the local level. This is largely because the court can only examine the application of domestic law, which does not effectively recognize the territorial rights of the Sámi people to lands nor protect their traditional livelihoods. As a result, the decisions of the court buttress national aspirations for sustainable development but depreciate the consequences of its implementation, including the specific risks this policy has for ensuring the sustainability of the traditional Sámi livelihoods. In fact, although the court decisions are symptomatic of the obligation for the court to examine and apply domestic law, the decisions favor a majoritarian market-oriented perspective of sustainable development which complies with national law but largely bypasses the rights of the Sámi Indigenous minority to ecological and cultural preservation at the local level as recognized under human rights law.

Seen through this lens, an important limitation of the court judgment thus lies in its failure to fully consider the status and rights of the Sámi reindeer herders as an Indigenous people recognized under international human rights law (Cambou, 2020). The court decisions are in fact the result of the application by the court of a legal framework which does not acknowledge the right of the Sámi as an Indigenous people, including their territorial rights, and which is therefore in breach of their human rights. In other words, the court decisions are confined by the assumption that the Swedish policy and legal framework comply with the rights of the Sámi as an Indigenous people. According to the court, "Swedish law is in accordance with its obligations/commitments under international law," in particular Article 27 of the ICCPR (Pauträsk and Norrbäck, 2019). However, this assumption can certainly be challenged considering Sweden's lack of compliance with a duty to respect and protect the rights of the Sámi in accordance with international human rights law (Cambou 2020).

Several international reports, academic contributions and recent court decisions (Girjas case, 2020; CERD, 2020) have questioned and challenged Sweden's commitments to the human rights of the Sámi as an Indigenous people. In its decision in 2020 regarding the adjacent Rönnbäck mining case, the UN Committee on the Elimination of Racial Discrimination (CERD) even concluded that Sweden had violated the right of the Vapsten *sameby* to property as enshrined in Article 5 (d) (v) of the Convention, notably because

> it has not complied with its international obligations to protect the Vapsten Sami reindeer herding community against racial discrimination by adequately or effectively consulting the community in the granting of the concessions.
>
> (CERD, 2020, para. 6.12)

This violation is a consequence of the lack of state engagement with the Sámi in relation to the governance and development of their land and natural resources

(Brännström, 2020; Allard, 2018; Larsen, 2019; Tauli-Corpuz, 2014). This situation is also similar to the wind energy context where governing institutions have continuously failed to recognize the right of the Sámi as an Indigenous people, treating them only as stakeholders and "industry," and neglecting their unique position as rights holders when making land use decisions (Larsen and Raitio, 2019; Cambou, 2020).

Intriguingly, a majoritarian and market-oriented understanding of sustainable development aimed at ensuring the sustainable growth of the Swedish economy is in fact applied to reindeer husbandry in so far as the court often refers to reindeer husbandry as an "industry." However, by weighing the interests of one industry against another, the court overlooks the importance of ensuring the maintenance of the reindeer "industry" in a way that is viable not only in the market sense but also culturally and ecologically, as established in human rights law. It is also notable that compensation provided to feed reindeer, support their transport by trucks and for the construction of barriers and corrals are unsatisfactory remedies, as they do not enable the practice of reindeer husbandry in a way that ensures animal welfare and meets Sámi economic and cultural needs (Tryland et al., 2019; Milner et al., 2014). These compensations offer a temporary "solution" that cannot account for the aggregated impacts caused by land fragmentation and the overall loss of grazing areas. In sum, the remedies offered by the court provide a quick fix to a situation that risks eroding Sámi traditional livelihoods in the long term and violates their human rights.

The responsibility of the court to protect the right of the Sámi as an Indigenous people is constrained by the dualistic approach of the Swedish system (Bogdan, 1994), which prevents the court from fully taking into consideration the human rights of the Sámi people in its decision-making. As a result, by seeking to ensure co-existence of wind energy development and reindeer husbandry in order to promote sustainable development, the court supports a governance structure that jeopardizes Sámi culture and their human rights. The judicial system does not confront the political imbalance between the two: national environmental and economic interests trumping the interests of the Sámi as an Indigenous people and the sustainability of their traditional lands and livelihoods at the local level. It may be beyond the scope of this analysis to question the role of the court in applying human rights law, but it can nonetheless be concluded that the court decisions are clearly limited in their ability to provide an adequate response for protecting the rights of Sámi reindeer herders as an Indigenous people. Equally, these conclusions also call into question the application of the principle of sustainable development insofar as it does not support a just transition for all that also respects the human rights of the Sámi people.

Conclusion

There is a paradox in the development of renewable energy: the objective to promote sustainable development may also lead to a situation where renewable energy could compromise sustainability. This paradox is vividly illustrated by the conflicts between wind energy developers and *samebyar* as discussed through the analyses of the court decisions in the Norrbäck and Pauträsk cases. In Sweden, the development of wind energy raises important governance challenges especially in light of the rapidly increasing pressures from other industrial developments on reindeer grazing land during the last decades. Now, wind energy presents itself as an additional challenge. Our analysis demonstrates how the promotion of wind energy in Sweden as a means to promoting sustainable development at the national level could jeopardize local sustainability by undermining the protection of the right of Indigenous Sámi reindeer herders to maintain their traditional livelihoods.

In the governance triangle, the Swedish court system has become an important arena for addressing these issues. However, the findings of our analysis stress the limitations of the court as an inadequate mediator to solve this conflict and ensure the protection of the Sámi right to conduct reindeer husbandry. The court decisions in the Norrbäck and Pauträsk cases demonstrate that there are several uncertainties in the balancing of interests between the goal to promote wind energy and the objective to protect and maintain reindeer husbandry. Some of these uncertainties lie in the difficulties of the court to conduct adequate assessments of the negative effects that wind energy projects can have on reindeer husbandry, specifically due to the lack of consensus about the way scientific results should be interpreted and regarding the scale and content of this assessment. The courts also appear to struggle with how to weigh information provided by true knowledge holders—the Sámi reindeer herders—about reindeer husbandry against the "knowledge" about reindeer husbandry provided by wind power companies.

Beyond the assessment issue, another shortcoming in the court decisions concerns the application of the principle of sustainable development as a meta-governance tenet guiding the court's decision-making. An analysis of the court rulings indicates that the decisions are framed and restrained by a legal and policy framework that favors national sustainable development but undermines the sustainability of Sámi reindeer husbandry at the local level. This is the result of the endorsement of domestic law which fails to fully recognize the right of the Sámi as an Indigenous people and which is therefore in breach of their human rights. In the absence of adequate recognition of the status and rights of the Sámi reindeer herders as an Indigenous people in Sweden, the conclusion of this analysis calls into question whether the court can successfully ensure a fair balance between the national goal to promote green energy and the rights of Sámi reindeer herders to conduct reindeer husbandry within their *sameby*. Ultimately, the decisions in the Norrbäck and Pauträsk cases epitomize the persistent challenges faced by the Swedish courts to guarantee sustainability at all levels in the face of increasing demands to promote sustainable development for all.

References

Allard, C. (2018) "The rationale for the duty to consult Indigenous peoples: comparative reflections from Nordic and Canadian legal contexts," *Arctic Review*, 9, pp. 25–43. Available at: https://doi.org/10.23865/arctic.v9.729

Anaya, J. (2011) *Report of the Special Rapporteur on the rights of indigenous peoples, James Anaya. Addendum: The situation of the Sámi people in the Sápmi region of Norway, Sweden and Finland.* A/HRC/18/35/Add.2. United Nations General Assembly, Human Rights Council.

Ava Case, No. M 10984-16 (2018) Svea Hovrätt Mark- och miljööverdomstolen [Svea Court of Appeal, Land and Environmental Court of Appeal]. Available at: https://www. domstol.se/mark--och-miljooverdomstolen/mark--och-miljooverdomstolens-avgoranden/2018/67492/

Bogdan, M. (1994) "Application of public international law by Swedish courts," *Nordic Journal of International Law*, 63, pp. 3–16. Available at: https://doi.org/10.1163/157181094X00734

Brännström, M. (2020) "The Girjas case—court proceedings as a strategy to enforce Sámi land rights," in T. Koivurova, E.G. Broderstad, D. Cambou, D. Dorough, and F. Stammler (eds.) *Routledge handbook of Indigenous peoples in the Arctic*. London: Routledge, pp. 174–186.

Cambou, D. (2020) "Uncovering injustices in the green transition: Sámi rights in the development of wind energy in Sweden," *Arctic Review*, 11, pp. 310–333. Available at: https:// doi.org/10.23865/arctic.v11.2293

CERD, Committee on the Elimination of Racial Discrimination (2018) *Concluding observations on the combined twenty-second and twenty-third periodic reports of Sweden*. CERD/C/ SWE/CO/22-23. United Nations International Convention on the Elimination of All Forms of Racial Discrimination.

CERD, Committee on the Elimination of Racial Discrimination (2020) *Opinion adopted by the Committee under article 14 of the Convention concerning communication No. 54/2013*. CERD/ C/102/D/54/2013. United Nations International Convention on the Elimination of All Forms of Racial Discrimination.

Forbes, B.C., Kumpula, T., Meschtyb, N., Laptander, R., Macias-Fauria, M., Zetterberg, P., Verdonen, M., Skarin, A., Kim, K.-Y., Boisvert, L.N., Stroeve, J.C. and Bartsch, A. (2016) "Sea ice, rain-on-snow and tundra reindeer nomadism in Arctic Russia," *Biology Letters*, 12(11), 20160466.

Girjas Case, No. T 853-18 (2020) Supreme Court of Sweden, January 23. Available at: https://www.domstol.se/hogsta-domstolen/avgoranden/2020/47294/

Ivan Kitok *vs.* Sweden (1988) Human Rights Committee, Communication No. 197/1985; UN Doc. CCPR/C/33/D/197/1985.

Klein, D.R. (1971) "Reaction of reindeer to obstructions and disturbances," *Science*, 173, pp. 393–398. Available at: https://doi.org/10.1126/science.173.3995.393

Lantto, P. and Mörkenstam, U. (2008) "Sámi rights and Sámi challenges: the modernization process and the Swedish Sámi movement, 1886–2006," *Scandinavian Journal of History*, 33, pp. 26–51.

Larsen, R.K. and Raitio, K. (2019) "Implementing the state duty to consult in land and resource decisions: perspectives from Sámi communities and Swedish state officials," *Arctic Review on Law and Politics*, 10, pp. 4–23. Available at: https://doi.org/10.23865/arctic. v10.1323

Larsen, R.K., Raitio, K., Sandström, P., Skarin, A., Stinnerbom, M., Wik-Karlsson, J., Sandström, S., Österlin, C. and Buhot, Y. (2016) *Kumulativa effekter av exploateringar på renskötseln: vad behöver göras inom tillståndsprocesser* [Cumulative impacts of land and natural resource exploitation on reindeer husbandry: what needs to be done in permit processes]. Rapport 6722. Stockholm: Naturvårdsverket.

Larsson, S. and Emmelin, L. (2016) "Objectively best or most acceptable Expert and lay knowledge in Swedish wind power permit processes," *Journal of Environmental Planning and Management*, 59, pp. 1360–1376. Available at: https://doi.org/10.1080/09640568. 2015.1076383

Lawrence, R. (2014) "Internal colonisation and Indigenous resource sovereignty: wind power developments on traditional Saami lands," *Environment and Planning D: Society and Space*, 32(6), pp. 1036–1053. Available at: https://doi.org/10.1068/d9012

Löfmarck, E. and Lidskog, R. (2019) "Coping with fragmentation. On the role of techno-scientific knowledge within the Sámi community," *Society & Natural Resources*, 32(11), pp. 1293–1311. Available at: https://doi.org/10.1080/08941920.2019.1633449

Milner, J.M., Beest, F.M.V., Schmidt, K.T., Brook, R.K. and Storaas, T. (2014) "To feed or not to feed? Evidence of the intended and unintended effects of feeding wild ungulates," *The Journal of Wildlife Management*, 78(8), pp. 1322–1334. Available at: https://doi.org/10.1002/jwmg.798

Norrbäck Case, No. M3053-15 (2017) (Umeå Mark- och miljödomstolen [Umeå Land and Environmental Court]). In file with the authors.

Norrbäck Case, No. M6974-17 (2019) (Svea Hovrätt Mark- och miljööverdomstolen [Svea Court of Appeal; Land and Environmental Court of Appeal]). Available at: https://www.domstol.se/mark--och-miljooverdomstolen/mark--och-miljooverdomstolens-avgoranden/2019/65429/

Omma, I.-A. (2021) Lawyer of the Vapsten *sameby*, personal correspondence with the authors.

Össbo, Å. and Lantto, P. (2011) "Colonial tutelage and industrial colonialism: reindeer husbandry and early 20th-century hydroelectric development in Sweden," *Scandinavian Journal of History*, 36(3), pp. 324–348. Available at: https://doi.org/10.1080/03468755.2011.580077

Pauträsk Case, No. M 3051-15 (2017) (Umeå Mark- och miljödomstolen [Umeå Land and Environmental Court]). In file with the authors.

Pauträsk Case, No. M 6860-17 (2019) (Svea Hovrätt Mark- och miljööverdomstolen [Svea Court of Appeal; Land and Environmental Court of Appeal]). Available at: https://www.domstol.se/mark--och-miljooverdomstolen/mark--och-miljooverdomstolens-avgoranden/2019/65441/

Pettersson, M. (2008) *Renewable energy development and the function of law: a comparative study of legal rules related to the planning, installation and operation of windmills*. Doctoral dissertation. Luleå: Department of Business Administration and Social Sciences, Luleå University of Technology.

Pettersson, M., Ek, K., Söderholm, K. and Söderholm, P. (2010) "Wind power planning and permitting: comparative perspectives from the Nordic countries," *Renewable and Sustainable Energy Reviews*, 14, pp. 3116–3123. Available at: https://doi.org/10.1016/j.rser.2010.07.008.

Prop. 1976/77: 80. *Regeringens proposition 1976/77:80 om insatser för samerna* [Government bill 1976/77:80 on initiatives for the Sami].

Rasmus, R. and Raitio, K. (2019) "Implementing the state duty to consult in land and resource decisions: perspectives from Sámi communities and Swedish state officials," *Arctic Review on Law and Politics*, 10, pp. 4–23. Available at: https://doi.org/10.23865/arctic.v10.1323

Sandström, P. (2015) *A toolbox for co-production of knowledge and improved land use dialogues— The perspective of reindeer husbandry*. Doctoral dissertation. Umeå: Swedish University of Agricultural Sciences.

Sandström, P., Cory, N., Svensson, J., Hedenås, H., Jougda, L. and Brochert, N. (2016) "On the decline of ground lichen forests in the Swedish boreal landscape: implications for reindeer husbandry and sustainable forest management," *Ambio*, 45(4), pp. 415–429.

SFS 1971:437, Reindeer Husbandry Act.

SFS 1974:152 (as Amended to December 7, 2010), Swedish Constitution.

SFS 1998:808, Swedish Environmental Code.

SOU 2020:73, Stärkt äganderätt, flexibla skyddsformer och naturvård i skogen [Strengthened property rights, flexible forms of protection and nature conservation in the forest].

Skarin, A. and Åhman, B. (2014) "Do human activity and infrastructure disturb domesticated reindeer? The need for the reindeer's perspective," *Polar Biology*, 37(7), pp. 1041–1054. Available at: https://doi.org/10.1007/s00300-014-1499-5

Skarin, A. and Alam, M. (2017) "Reindeer habitat use in relation to two small wind farms, during preconstruction, construction, and operation," *Ecology and Evolution*, 7, pp. 3870–3882. Available at: https://doi.org/10.1002/ece3.2941

Skarin, A., Nellemann, C., Rönnegård, L., Sandström, P. and Lundqvist, H. (2015) "Wind farm construction impacts reindeer migration and movement corridors," *Landscape Ecology*, 30, pp. 1527–1540. Available at: https://doi.org/10.1007/s10980-015-0210-8

Skarin, A., Sandström, P., Alam, M., Buhot, Y. and Nellemann, C. (2016) *Renar och vindkraft II – Vindkraft i drift och effekter på renar och renskötsel* [Reindeer and wind power II—Wind power in operation and effects on reindeer and reindeer husbandry]. Rapport 294. Uppsala: Sveriges lantbruksuniversitet SLU (Swedish University of Agricultural Sciences).

Skarin A., Sandström P. and Alam M. (2018) "Out of sight of wind turbines—reindeer response to wind farms in operation," *Ecology and Evolution*, 8(19), pp. 9906–9919.

Sköld, P. (2015) "Perpetual adaption? Challenges for the Sámi and reindeer husbandry in Sweden" in B. Evengård, J. Nymand Larsen, and Ø. Paasche (eds.) *The new Arctic*. Cham: Springer International Publishing, pp. 39–55.

Solbär, L., Marcianó, P. and Pettersson, M. (2019) "Land-use planning and designated national interests in Sweden: Arctic perspectives on landscape multifunctionality," *Journal of Environmental Planning and Management*, 62, pp. 2145–2165. Available at: https://doi.org/10.1080/09640568.2018.1535430

Tauli-Corpuz, V. (2014) *Report of the Special Rapporteur on the rights of indigenous peoples*. A/HRC/27/52. United Nations General Assembly, Human Rights Council.

Tryland, M., Nymo, I.H., Sánchez Romano, J., Mørk, T., Klein, J. and Rockström, U. (2019) "Infectious disease outbreak associated with supplementary feeding of semi-domesticated reindeer," *Frontiers in Veterinary Science*, 6, article 126. Available at: https://doi.org/10.3389/fvets.2019.00126

Vistnes, I. and Nellemann, C. (2008) "The matter of spatial and temporal scales: a review of reindeer and caribou response to human activity," *Polar Biology*, 31, pp. 399–407.

4 Indigenous agency in aquaculture development in Norway and New Zealand

Camilla Brattland, Else Grete Broderstad and Catherine Howlett

Introduction

Indigenous peoples are one of the marginalized groups in developing countries—along with women, minorities and small-scale fishers—that experience unequal distribution of ocean resources and benefits. In developed nations such as Norway and New Zealand, Indigenous peoples have greater access to many of the key assets to achieve equity, including education, political organizations and physical assets that other Indigenous people may experience barriers in achieving. Nonetheless, inequality in ocean governance is also a reality even in these seemingly resourceful contexts (Hersoug 2005; Jentoft and Eide 2011; Brattland 2013; Hersoug 2018), making the capacity of Indigenous groups to participate in and influence marine development a relevant aspect of ocean equity.

In this chapter, we discuss Indigenous political agency in aquaculture industry development in two contexts, Norway and Aotearoa–New Zealand (NZ), and understand Indigenous political agency as the power to act (Tennberg 2010). In the interaction between Indigenous peoples and a global market player such as the aquaculture industry, the agency of Indigenous peoples is mediated by a variety of discursive, legal and institutional realities existing within different national jurisdictions. The global salmon industry is led by multinational companies, many of which originate from or have their head offices in Norway. Although there are also smaller and family-owned aquaculture companies in Norway, such as those in the western parts of the country, this is not the case in Sámi coastal areas, where ownership is dominated by the multinational companies MOWI, Grieg Seafood and Cermaq. In NZ, too, the industry operates in territories where Indigenous peoples are affected. Salmon farming in general is subject to international standards on Indigenous rights as well as consumer and public expectations on environmental sustainability and social responsibility, including respect for Indigenous rights and territories. Indigenous peoples' power to act is thus influenced by the multiple forces of economic development pressures from the global aquaculture industry and the Indigenous rights structures and discourses aimed at preserving Indigenous cultures and livelihoods.

Indigenous rights to marine resources used to be unrecognized as the oceans were traditionally conceived of as "mare nullius," which excludes Indigenous peoples as marine rights holders (Mulrennan and Scott 2000). According to legal scholars

DOI: 10.4324/9781003131274-4

Allen et al. (2019), claims raised by Indigenous peoples on the right to marine resources have been largely opposed. While Norway emphasizes legislation and politics on Sámi rights as an obligation under international law stemming from various human rights conventions, New Zealand's law and politics is informed by the duty of the Crown as recognized under the nation's founding document, the Treaty of Waitangi. Legal structures thus both enable and constrain Sámi and Māori agency in the two countries, where Indigenous agency in NZ appears stronger than in Norway.

The Māori of Aotearoa–New Zealand are often looked to as an example for other Indigenous peoples to follow, including the Sámi. An important point of departure for this chapter is the unequal resource and property rights recognition in relation to aquaculture development, which from the outset enhances Māori agency in marine development. Yet, legal status and recognition is not a guarantee for agency in the interaction between Indigenous peoples and industry. From a political ecology standpoint, political scientist Catherine Howlett (2010) conceptualizes Indigenous agency as constrained by multiple factors such as historical and discursive relations, which cannot be solved through legal recognition and political negotiations alone. In the case of mineral conflicts in Australia, she finds that the agency of the state and the mining company is ultimately favored at the expense of Indigenous agency, due to the capitalist structural, material and ideational reality that has informed mineral development (Howlett 2010, p. 117). The aquaculture industry is similarly driven by a global market demand that now favors the use of marine space for farmed salmon production rather than for Indigenous marine livelihoods. Seafood companies are key drivers and players in the globalization of farmed salmon as a prized seafood commodity. In Canada, Chile, Norway and New Zealand, the production of salmon takes place in Indigenous marine territories that had value and significance for Indigenous communities before the advent of the industry. What are the main factors, then, influencing Indigenous peoples' power to act for equity in aquaculture development?

Our intention in this chapter is to arrive at some recommendations for how Indigenous agency in aquaculture development can be strengthened. We identify and discuss structural and discursive factors, where structural factors can include governance instruments stemming from Indigenous rights recognition, and where discursive factors include pervasive discourses that may influence the agency of the actors involved. Against this background, our contribution is to 1) identify the main relevant structural and discursive factors that affect Indigenous political agency in aquaculture development in two different cases of Indigenous–industry interactions in northern Norway and New Zealand, and 2) arrive at some lessons and recommendations from each case that can enhance the agency of Indigenous peoples in aquaculture development.

The chapter consists of five sections. The introduction is followed by Section Two, which gives an account of our analytical approach by presenting the concepts of agency, structural and discursive constraints and enablers. Section Three starts with a presentation of the methods and data before proceeding with the NZ and Norwegian cases. In the fourth Section, the main structural and discursive factors influencing Indigenous agency are identified and discussed relative to each other, and we conclude with recommendations for enhancing agency in Section Five.

Indigenous political agency

To ensure social equity in governance of marine resources, it is important to recognize the rights and needs of all groups that depend on the ocean in decision-making as well as in benefit sharing (Österblom et al. 2020). This makes the agency of Indigenous peoples in coastal communities a central factor for achieving greater equity in ocean governance. Indigenous peoples in coastal communities may experience both procedural and distributional inequity, as the benefits from aquaculture development tend to reach a few private actors, and involvement in aquaculture production varies with state and private recognition of Indigenous peoples' resource rights (Österblom et al. 2020).

There is an increasing focus in academic research and literature upon Indigenous "agency" as a critical factor when examining the issue of resource governance on Indigenous lands and territories (see Trebeck 2007). Many scholars attribute much of the successful outcomes that Indigenous peoples have obtained in resource governance arrangements to Indigenous agency. Agency in this context refers to Indigenous peoples' ability to affect their environment with their political conduct or, similarly, their ability or capacity to act consciously to realize their intentions (McAnnulla 2002). Political scientist Monica Tennberg (2010) likewise defines Indigenous political agency as the power to act. This power may take different forms of political action, including self-identification, participation, influence and representation. Here we emphasize participation and influence, entailing the ability to act and participate meaningfully in policy- and decision-making on aquaculture development, which is not the same as having the final word or veto on equity in ocean governance. Indigenous peoples' political agency is a result of the operation of multiple power relations, which not only define the scope of political action and its forms, but also constrain and enable Indigenous political agency. Thus, securing Indigenous perspectives in mainstream decision-making bodies at local, regional and national levels becomes a prerequisite for safeguarding Indigenous peoples' ability to act and participate meaningfully, and to have their views included in policymaking (Broderstad 2014). In the case of Norway, this "breaking in approach" (Josefsen 2014) underlines the autonomy of the Sámi Parliament of Norway as the political representative for the Sámi, while Sámi concerns are simultaneously incorporated into legislation and decision-making through policymaking.

Numerous structural and discursive factors influence Indigenous agency to the extent that they constrain and enable Indigenous agency in resource governance processes, both nationally and internationally. Structural factors may include legislation and policies that inform resource governance in national jurisdictions. The colonial and historical distribution of resources and interests laid down structurally over time may also exert an influence on agency (Howlett 2010). At the international level, the development of international Indigenous activism in various collective structures has enhanced Indigenous agency (Tennberg 2010), resulting in milestone achievements of international law (Dahl 2012; Åhrén 2016).

Indigenous agency can also be influenced by the lack or quality of the knowledge basis for resource governance and market realities. That is, those groups with greater access to resources and knowledge have greater capacity to participate in resource decision-making processes on a more level playing field. This knowledge can be mediated by discursive factors, which determine what knowledge and what political claims are seen as possible or legitimate. They can also determine who can make authoritative claims to hold such knowledge or make those political claims (Reimerson 2016). Often in resource governance processes, Indigenous knowledge can be delegitimized or regarded as inferior to scientific knowledge. Indigenous agency in resource governance processes can thus be subject to discourses shaped and influenced by colonial histories and post-colonial power relations that may render Indigenous knowledge as inferior or as some form of folk knowledge. In the decision-making processes for aquaculture licensing, Indigenous knowledge may be pitted against scientific data, particularly if the scientific data supports aquaculture developments as beneficial and environmentally sustainable.

In the case of aquaculture development in Indigenous territories, there is an uneasy balance between the three sustainability dimensions—economic, social and environmental—as the industry may bring increased economic and social benefits for local communities but result in degradation of marine and anadromous fish species and ecosystems (Young et al. 2019). Development is thus more often discussed as a process that Indigenous peoples have the right to be protected from rather than participate in as active players (Blaser et al. 2004). The social sustainability dimension of industrial development has been criticized for being translated too narrowly into, for instance, the number of local employees (Eythórsson et al. 2019). Underlying this discourse on sustainability and development is the extent to which Indigenous rural communities find themselves in a state of "underdevelopment," resulting in increased dependence on economic development either by state actors or global multinational companies.

Indigenous agency in aquaculture development in Norway and New Zealand

Norway and New Zealand both have Indigenous peoples with strong historical and cultural ties to the sea and marine resources who have raised claims of rights to these marine resources and the right to participate in their management. In NZ, Indigenous marine rights recognition is advanced compared to Sámi historical marine rights recognition in Norway which is lacking. This does not necessarily mean that Māori interests are always heard and respected in aquaculture licensing processes, nor does it mean that Sámi are without capacity to influence marine governance and aquaculture development in particular. The two cases that we discuss in relation to each other are cases of Indigenous presence and involvement in aquaculture development in the Marlborough Sounds and in the western Finnmark region. Both salmon aquaculture and Indigenous presence is, however, vastly different between the two contexts. The intention behind the selection of the case studies was to contrast Sámi agency in aquaculture development, or the lack thereof, with a case where Indigenous peoples were recognized as major players in the seafood sector. In this respect, the Māori of Aotearoa–New Zealand seem exceptional.

In terms of the volume of the industry in the two contexts at the national level, there are more differences than similarities. The production volume of farmed salmon in New Zealand is only a fraction of what Norway produces (in 2019, New Zealand exported around 5,000 tons of king salmon against Norway's 1,200,000 tons of Atlantic salmon), and New Zealand's main farmed produce are shellfish (Greenshell mussels) rather than finfish. We focus our analysis on two cases set in the larger regions of western Finnmark and the Marlborough region, home to around 50,000 people each (Statistics Norway 2021; Statistics New Zealand 2021). Within these regions, salmon farming takes place in the coastal part of the western Finnmark region and the Marlborough Sounds, which is part of the *iwi* territory of Ngāti Toa Rangatira (see Figure 4.1). (*Iwi* is the name given to the Māori tribes of Aotearoa). The coastal part of the western Finnmark region is approximately the same size as the Marlborough Sounds area, has around the same number of salmon farms (see Figure 4.2), and both regions are predominantly rural with most inhabitants settled in small towns and urban centers (Honningsvåg in Norway and Picton in NZ). Whereas the proportion of Māori in the Marlborough district is known (13.3 percent in 2020), the proportion of Sámi people in Finnmark is unknown, as population censuses do not register ethnicity. As a rule, however, the coastal areas are inhabited with a minority of Sámi, while the inhabitants of the inland or "core" areas are predominantly of Sámi descent. A majority of the municipalities in the Troms and Finnmark County are also part of the Sámi

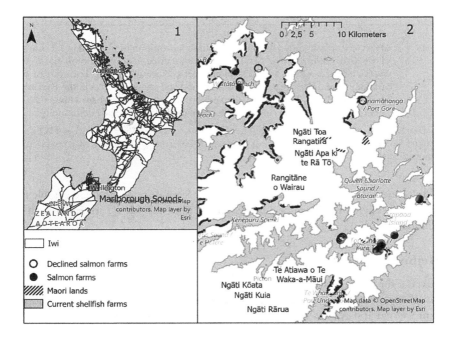

Figure 4.1 Overview of case study area and aquaculture locations in New Zealand. Map 1. Borders of *iwi* in New Zealand and location of Marlborough Sounds. Map 2. Locations of shellfish farms and current and declined salmon farms in Marlborough Sounds, and names of affected *iwi* in the area.

Source: LINZ data service: https://data.linz.govt.nz/ (Accessed March 15, 2021) and Open Street Map. Map produced by Camilla Brattland.

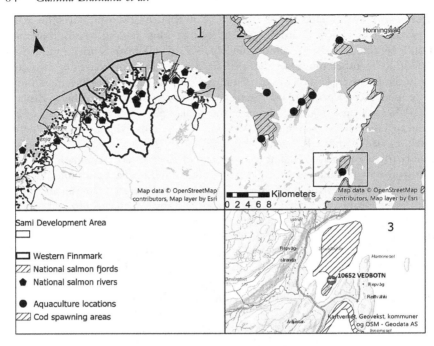

Figure 4.2 Overview of case study area and aquaculture locations in Norway. Map 1. Location of western Finnmark, the Sámi Development Area, national salmon fjords and rivers and aquaculture locations in northern Norway. Map 2. Location of the Stranda community, aquaculture locations and cod spawning areas in the Porsanger Fjord area. Map 3. Location of the Vedbotn salmon farm, cod spawning areas in Vedbotn/Stránddavuotna and salmon rivers used by the Stranda community.

Sources: Norwegian Directorate of Fisheries, Norwegian Directorate for Water and Energe (NVE), and Statistics Norway. Basemap from Open Street Map and © Kartverket (CC BY 4.0). Map produced by Camilla Brattland.

Development Area, whose inhabitants (whether Sámi or non-Sámi) are eligible for economic support for livelihood development from the Sámi Parliament such as funding for fishing vessels.

Methods and data

Material for the case studies was predominantly gathered between 2016 and 2020, based on a qualitative case study design with field visits to the case study locations with known Indigenous controversies over salmon farming, and interviews with key actors. The field visits and interviews were preceded by a literature review and desktop study of key figures and statistics on Indigenous populations and salmon farming in the two regions, as well as by gathering previous research on the topic. The same interview guide was used for both contexts in interviews conducted in English in New Zealand and Norwegian in Norway. For the Norwegian case, Else Grete Broderstad interviewed an industry representative (Lise Bergan, Corporate

Affairs Director, Cermaq), while Camilla Brattland interviewed a member of the Norwegian Sámi Parliament Council, responsible for fisheries and coastal affairs (Silje Karine Muotka). Brattland also visited the Finnmark region and the Porsanger and Nordkapp municipalities in the summer of 2019, interviewing three local inhabitants in the Stranda community, using a map-based biography method to document historical land use and the associated nature values. Catherine Howlett produced a case study on Māori relations with the aquaculture industry in New Zealand, based on fieldwork conducted in the spring of 2018, interviewing both involved *iwi* Māori and representatives of The New Zealand King Salmon Co. Ltd. She conducted a total of seven interviews with industry and Māori representatives. The interviews and field visits in the western Finnmark region and the Marlborough region were developed into case narratives (Yin 2012; Bartlett and Vavrus 2016) that focus on the involvement of Sámi and Māori in licensing processes. They are not intended as a full documentation of the complexity of multiple dimensions of aquaculture development and Indigenous agency in the two regions. Statistics and data sets from public authorities relevant to the case studies were collated for Figures 4.1 and 4.2. The most striking difference between the two contexts is the lack of Sámi presence as legal actors in the Finnmark region as opposed to the clear recognition of Māori as legal actors in everything from detailed statistics to demarcated areas for Māori property ownership. In the next sections, we present the case studies of Māori and Sámi involvement in aquaculture development.

Māori marine resource rights context in Aotearoa–New Zealand

In global Indigenous rights discourse, Aotearoa–New Zealand is often presented as "leading the way" or "setting the agenda" in terms of recognition of Māori rights, particularly in relation to resource governance and management (see Bargh 2018). In relation to fisheries management, NZ has been twice ranked as the most sustainable in the world (McCormack 2017, p. 36). This view is not uncontested, particularly by Māori scholars, or scholars researching within NZ (see, for example, Bell 2018; McCormack 2017).

Prior to colonization, Māori had developed a sophisticated management system, with detailed rules to regulate who could fish for what, where and when, and were involved in trade and barter long before British arrival. The general principle was that each tribe (*iwi*) or subtribe (*hapu*) controlled the waters adjacent to their land, areas which were demarcated in detail and where trespassing would be punished (Hersoug 2018, p. 104). Thus, all land in NZ was once Māori customary land (held by Māori in accordance with *tikanga* Māori—Māori laws and customs) (Ruru 2010, p. 67). The treaty simply gave the British Crown the right to govern Māori (Ruru 2010, pp. 5–7). One of the resources seemingly and specifically reserved for Māori under the treaty was fisheries (Bargh 2016, p. 2). According to Hersoug (2018, p.102), under the treaty, Māori were ensured a guarantee from the British Queen that they should keep "the full, exclusive and undisturbed possession of their fisheries"—a claim that has never been rescinded in later legislation.

The Resource Management Act (RMA) of 1991 is the primary legislation which governs the establishment of marine farms. Anyone wishing to establish a marine farm must apply for and obtain a resource consent from the appropriate regional council or unitary authority. The Māori Commercial Aquaculture Claims Settlement Act 2004 (Settlement Act) (amended in 2011) was enacted to provide *iwi* with access to aquaculture space to develop their marine farming interests or obtain cash settlements if appropriate. The settlement required the government to provide *iwi* aquaculture organizations with settlement assets representative of twenty percent of all new aquaculture management areas created in coastal waters since 1992 and twenty percent of any new marine farming space. In addition to economic interests (and often in opposition to them), Māori hold traditional ecological knowledge that is based on customary rights and cultural traditions that support the multiple values carried by coastal marine areas. In particular, their role, duties and rights under the principle of *Kaitiakitanga*—as stewards of the environment—means that they have the right not only to participate in management but also manage resources in NZ. *Kaitiakitanga* literally means guardianship and protection. It is a way of managing the environment, based on the Māori worldview that humans are not separate, but part of the natural world. It is the local version of more internationally recognized traditional environmental knowledge and operates locally within communities and within specific environments to ensure maintenance and sustainability of the resources in that region (McGinnis and Collins 2013, pp. 412–416). In NZ companies must consult with communities or reach agreements with the relevant Indigenous governance bodies who are in a position to consult with. These are the relevant *iwi* tribal organization and the Te Ohu Kaimoana (TOKM), which was established through the Māori Fisheries Act 2004 to act as trustee with responsibility for the allocation of Māori Fisheries Settlement assets to *iwi*.

Aquaculture industry development in Te Tau Ihu, Marlborough Sounds

Most of the farmed salmon in NZ is produced in Marlborough Sounds, which has the best water conditions for farming. New Zealand accounts for over half of the world production of king (or Chinook) salmon (*Oncorhynchus tshawytscha*) (Ministry of Primary Industries [MPI], 2017), which is the only salmon species produced in the country. The New Zealand King Salmon Company (NZKS) is the primary producer of king salmon (McGinnis and Collins 2013, p. 401) and employs 480 people in total (as of 2020), whereas less than 140 are registered as employees in the agricultural and fisheries sector in the region (Statistics New Zealand 2021). The total number of marine farms in NZ is 1,508, whereas salmon farms constitute fewer than twenty farms either as salmon-only licenses or licenses that combine shellfish, salmon and algae.

A recent proposal to relocate salmon farms in the Marlborough Sounds provides a case study to examine the influences acting upon Māori agency in aquaculture developments. Marlborough Sounds is a sensitive and unique coastal marine area, which covers some 4,000 square kilometers of sounds, islands and peninsulas at the South Island's north-easternmost point (McGinnis and Collins 2013, p. 402). It is an area where there have been several conflicts over aquaculture in the past, particularly as a result of Māori claims of customary rights to the foreshore and seabed in the area (Hersoug 2002). There are eight *iwi* involved in the negotiations over the current proposal in Te Tau Ihu (Nelson–Marlborough). There are *iwi* who support the proposal and seek employment and income opportunities from aquaculture development in the Marlborough Sounds, hoping to benefit materially from the twenty-percent allocation under the Māori Commercial Aquaculture Claims Settlement Amendment Act 2011, and there are *iwi* who resist any salmon farming in their traditional waters and feel their knowledge and perspectives are marginalized. In October 2011, The New Zealand King Salmon Company applied to the Environmental Protection Authority for consent to relocate nine salmon farms in the Marlborough Sounds. Eight *iwi* may claim customary interests where the farms were proposed, and thus any expansion would be significant in terms of the Treaty of Waitangi (Inns, 2018a; Minister for Primary Industries (MPI) 2017). Māori interest in aquaculture is growing in part as a result of claims from Treaty of Waitangi Settlements and from growing joint venture agreements. The allocation of twenty percent of new aquaculture space under The Māori Commercial Aquaculture Claims Settlement Amendment Act in 2011 means Māori are key economic stakeholders in the allocation of coastal marine space (McGinnis and Collins 2013, pp. 412–416).

Those *iwi* opposed to the proposals to relocate the salmon farms have cited many reasons for the opposition, including the incompatibility of salmon farms with their custodial responsibilities and cultural aspirations:

> They don't fully understand *iwi*'s connection to the waterways and *kaitiaki* that look after those waterways' deep spiritual connection [the waterways have both a spiritual and a present connection] between *iwi* and their environments—had no respect for other Indigenous people in other areas where they have commercialized the treasures of Indigenous people….we are opposed to destructive, environmental practices condensed in a small and uniquely valued, pristine waterway of which we are *kaitiaki*.
>
> (Interview, April 2, 2018b)

The *iwi* opposition to the relocation of the salmon farms was rooted in their deeply held environmental and cultural knowledge of their lands and sea, and many *iwi* were offended by the treatment of their cultural and environmental knowledge. There was explicit reference to the hegemonic discourse of scientific knowledge as superior to Indigenous knowledge. This was reflected in comments made by the CEO of King Salmon in newspaper articles at the time, which is detailed below.

Those *iwi* who were not opposed to the proposal to relocate the salmon farm and increase salmon farm production in the Marlborough Sounds—the Te Ātiawa—argued that they had to be involved in any salmon farming activity on that site, since they held a coastal permit for previous aquaculture activities in one of the proposed designated areas. In an interview with the Chairperson of Te Ātiawa o Te Waka a Māui Trust, the Chairperson stated that Te Ātiawa are not opposed to developments in their *rohe* (geographical and spiritually bounded territories and waters) which bring benefits to the people, as long as they are not in breach of the *kaitiaki* for their *rohe*. He stressed that Te Ātiawa have obligations as *Kaitiakitanga* or guardians of their environment and that if business opportunities did not breach those obligations, then he and his *iwi* were not opposed to engaging with business (Interview, April 16, 2018e). According to the Aquaculture Stewardship Council (ASC), the Clay Point farm in the Marlborough Sounds is certified under the ASC sustainability and labeling program. The farm had no complaints from Māori, and the company has a protocol agreement and have undertaken consultations with the relevant Indigenous governance bodies in the area (ASC 2020b).

When a panel of environmental experts concerning the relation and expansion of salmon farming in the Marlborough Sounds recommended the decision that the future for salmon farming must lie in land-based or open-sea farming, NZKS representatives were adamant that the Crown would "bend over backwards to accommodate *iwi*" and clearly believed that in NZ the Crown acts primarily in the interests of *iwi* and does not give any preferential treatment to the industry.

The *iwi* of Ngāti Kuia has had a long historical association with the waters of the Marlborough Sounds and have some of the oldest known oral traditions and histories of Māori. Several submissions were made to the environmental experts' panel by members of this *iwi*, including Māori historian Peter Meihana, and Raymond Smith. Meihana explained that while critical European thinking may regard some Māori oral traditions as mingling time and myth, from the Māori point of view they are oral accounts sustaining the whole Māori system of *iwi* or *hapu* (Māori term for clan or descent groups). *Iwi* are made up of several *hapu* customary entitlements, responsibilities and obligations. Meihana expressed the oral traditions as setting the responsibilities and obligations for *hapu* (MPI 2017, p. 57). Meihana is worth quoting at length from his submission to the panel for the insights he provides regarding the treatment of Māori knowledge and oral traditions:

> Before proceeding it is important to consider that since the advent of British colonization, Māori oral tradition and the intellectual *whakapapa* that underpins and orders those traditions, that is to say *whakapapa*, have been subjected to European analysis. The impact of such analyses on Māori has been profound. Oral traditions were constructed within a particular spatial and temporal context. To separate space and time, the characteristic of western approaches to the past produces a history. To subject oral tradition to such treatment consigns it to the realms of myth and therefore without legitimacy. This has, of course, been central to colonization here and overseas. In New Zealand the taking of land was nearly always proceeded with attempts at undermining Māori views of reality. Consequently those peoples were deemed irrational without reason and could therefore be justifiably dispossessed.
>
> (Meihana 2017, pp. 27–28)

He offers an overview of *Whakapapa* in his submission, stating that this is not simply a family tree of vertical and horizontal familial relationships. *Whakapapa* is a system of knowledge, an epistemology that can explain the world around us and legitimates rights to resources (Meihana 2017). Raymond Smith of Nagti Kuia *iwi* also made a presentation/submission to the panel and offered the following insightful statement on the epistemological challenges inherent in many contestations concerning resource developments such as this aquaculture proposal:

> We try and understand science, it's really hard and I think I've seen some of the looks on your faces when I hear some of the science with all the technical data that flows through that in a totally different language, we have to understand. So there's three languages that we're trying to understand here: English, Māori and science. So our *matauranga*, our Indigenous *matauranga* is a science.
>
> (Smith 2017, p. 30)

He went on to argue the proposal was contrary to the cultural aspirations of Ngāti Kuia and concluded his submission with a plea to the panel "to look through our cultural lens, … to feel our cultural connection and our obligation to take your decision into the future as *kaitiaki* of our *rohe*, our history and our culture (Smith 2017, p. 32).

Ngāti Kuia presenters argued that the oral accounts of their relationship and *Kaitiakitanga* role with their *rohe* were often discounted by Europeans, who tend to dismiss the oral traditions as myths or as recent convenient constructs (MPI 2017, p. 57). The presenters asserted that the CEO of NZKS's Grant Rosewarne caused great offense to many Ngāti Kuia by questioning their belief in the "white dolphin" *Kaikaiawaro*, a key aspect of Ngāti Kuia identity and *kataiaki* or guardian of the area, inferring it was a recent convenient construct. Rosewarne publicly challenged the integrity of their knowledge in a media interview: "I doubt whether a majority hold that belief about the white dolphin Kaikaiawaro" (McPhee 2017). Thus, in this case study on how the Crown mediates the Māori–business relationship, who it privileges and who it subjugates, the views of some Māori and business are diametrically opposed.

Sámi marine resource rights context in Norway

The existence of Sámi historical fishing rights was documented by a government-appointed expert committee in 2006 tasked with investigating the rights of the Sámi and others to saltwater fishing outside Finnmark. Based on international law, the committee unanimously concluded that sea Sámi fishing rights exist, enjoyed by both Sámi and non-Sámi in Finnmark (Smith 2014, p. 5). Still, Sámi historical fishing and marine resource rights remain unrecognized by the Norwegian government. However, rights claims pertaining to delimited sea areas may be raised to the Finnmark Commission, which is responsible for documenting property ownership in the Finnmark region under the Finnmark Act (2005). The Marine Resource Act recognizes that due concern shall be paid to Sámi use of marine resources and to Sámi communities upon allocation of quotas and other management measures (Marine Resource Act (2009), § 11, final section). This does not, however, include aquaculture allocation licenses, which are governed by the Aquaculture Act (2006).

The most important legal instrument for the Sámi on aquaculture licensing is the role of the Sámi Parliament in municipal coastal zone planning. The Planning and Building Act (2008) obligates planning authorities to "protect the natural basis for Sami culture, economic activity and social life" and grants the Sámi

Parliament the authority to file formal objections to new plans "in respect of issues that are of significant importance to Sami culture or the conduct of commercial activities." Municipalities are obliged to address such objections, usually by modification of the plan in agreement with the Sámi Parliament. Impacts on the natural basis of Sámi culture and for coastal Sámi communities are among the issues to be assessed for each proposed location. Coastal zone planning under the Planning and Building Act is an important governance instrument to avoid or minimize conflicts over allocation of marine space to aquaculture. This is achieved by informed decisions, stakeholder participation and transparency (Hersoug and Johnsen 2012). Municipal authorities are required to plan in and around these sea areas before any allocation of aquaculture sites can take place. Suitable areas for salmon farming must be specified and mapped in the planning processes, which involves input from several government agencies and stakeholders. When a plan is approved by the municipal and regional authorities, aquaculture companies are not granted ownership to a site, but are allowed to pursue their operations provided they have a license, which is allocated for a specified period. The license can be withdrawn by the authorities and the sites can also be shut down upon outbreaks of disease.

The right to be consulted as another instrument on matters pertaining to the Sámi, is instituted in the Consultation Agreement (2005) between the Sámi Parliament and the Norwegian state based on Norway's ratification of the International Labour Organization's Indigenous and Tribal Peoples Convention no. 169 (ILO 169). The agreement gives the Sámi Parliament the right to be consulted in all relevant matters and is the core tool of interactions on central policies, including resource and fisheries management, with the Norwegian government and its ministries, directorates and other subordinate state agencies or activities (Procedures for consultations, point 2). However, the Norwegian National Human Rights Institution (NIM 2016) has noted that these arrangements are not sufficient to fulfill Norway's human rights obligations toward the Sámi in the marine resource rights field. Importantly, local communities do not have the same right as the Sámi Parliament to be consulted on issues that pertain to Sámi culture or livelihoods. This leaves a gap in the system which depends on the agency of local communities to raise issues with multiple authorities. A bill proposed in 2018, to make the right to consultations statutory by amending the Sámi Act, was adopted in June 2021. Municipalities and county municipalities are obliged to consult with Sámi communities on issues affecting them. A natural starting point at the municipal level will be to consider how local Sámi interests judge the need for consultations. An obligation to consult can, for instance, apply to allocation of aquaculture licenses if the measures impact sea Sámi communities and interests (Draft resolution and bill 2020–2021, p. 108). In the next section on the case study, the current consultation gap is dealt with as one of several factors influencing the agency of the Sámi in aquaculture development at the local level.

Aquaculture industry development in Sámi areas

Around the turn of the twenty-first century, the environmental effects of salmon farming were not on the agenda as much as the material benefits and employment opportunities that the farming brought for rural communities in coastal northern Norway. Sámi politicians argued that Sámi should have aquaculture licenses allocated to their territories for them to benefit from the industry. In an exceptional case, two licenses were allocated to the Lule Sámi community of Tysfjord to support the material basis for Sámi culture in the area through fish farming. The licenses were owned by local Sámi, but were later sold to larger aquaculture companies (Evjen 2014; Sandersen 2017). This was the only case of something like Māori ownership in the Sámi context, where local ownership was intended to increase Sámi employment and welfare. Since the Norwegian government abandoned priority for local ownership and distribution of licenses to rural and northern districts as governance instruments in the 1990s (Hovland 2014), there has been no special aquaculture development policy for Sámi areas.

With increasing global ownership and awareness of environmental impacts on the marine environment and genetic pollution of wild salmon species, the latter years have seen an increase in conflicts between small-scale fisheries and aquaculture locations in coastal Sámi areas. The Sámi Parliament argues that salmon farming impacts Sámi communities negatively in terms of direct environmental impacts on wild Atlantic salmon stocks and on marine species, which again affects the natural basis for Sámi culture and livelihoods.

Aquaculture policy is an area where the Sámi Parliament has little agency relative to its rather active role in national marine fisheries management, such as on the issue of cod quota allocations. The general attitude from the authorities seems to be that since Sámi did not subsist from aquaculture traditionally, there is no reason to consult on it as a relevant issue. A member of the Sámi Parliamentary Council argues that the Ministry of Trade, Industry and Fisheries has systematically rejected repeated requests to consult on national aquaculture development policies (Interview, June 25, 2018c). Among the development policies were the criteria for where and how much the industry should be allowed to grow in the different Norwegian regions (the so-called fivefold growth policy). Without real participation in ocean policy decision-making, the remaining important governance instruments in which the Sámi Parliament and local communities can participate are thus the Aquaculture Act and related instruments (Brattland 2013), the municipal and regional coastal zone planning processes, and sustainability principles for the industry at large, such as the ASC criteria.

In an interview, we asked a Cermaq representative where this major salmon farming company could encounter Sámi interests in marine areas. The reply: in cases between fish farming locations and reindeer migration routes crossing inlets, and in the placement of salmon abattoirs. Compared to Canada, where the situation is more conflictual and where the company also has Protocol Agreements with First Nations, the conflict level in Norway is considered lower due to the planning regime, which contributes to preventing conflicts. The company relates to communal plans and has clear rules of operations. Prior to drawing up a license application, the case must be properly documented and, in that context, potential conflicts with Sámi interests can be mapped (Interview, May 25, 2018a). According to the Sámi Parliament there is, however, no systematic mapping of coastal Sámi interests in fishing areas, nor is there an agreed-upon policy for protection of Sámi interests in aquaculture cases.

Salmon farm licensing in Porsanger Fjord, western Finnmark

A relevant example of how the lack of a coherent aquaculture policy for Sámi areas affects Sámi agency is the licensing process for the Vedbotn location in western Finnmark. In the Finnmark region, around 600 persons were employed in the salmon aquaculture industry in 2019. This is less than half of the 1,425 active fishers who owned fishing vessels in the same year. Around 120,000 tons of farmed salmon were sold from 106 farms located mostly in the southern part of the region. All of the coastal Finnmark district is recognized as a coastal Sámi settlement area (Marine Resource Act 2009, § 21), and the Porsanger Fjord area is home to Sámi settlements who have subsisted from marine resources long before the coast was settled by Norwegians (Brattland and Nilsen, 2011), and whose inhabitants also participate in commercial marine fishery (Brattland 2013; Broderstad and Eythórsson 2014). The Vedbotn location is placed close to the border between the Porsanger and Nordkapp municipalities, which have opposing views on aquaculture development. Whereas the Porsanger municipality hosts a national salmon fjord (see Map 1, Figure 4.2) with restrictions on the establishment of salmon farms due to the impacts of sea lice on wild salmon, the Nordkapp municipality is a historical center for fisheries industry development, which also includes aquaculture development. When the Nordkapp municipality board decided to approve Grieg Seafood's application to establish a salmon farm in Vedbotn, the inhabitants of the adjacent settlement of Stranda were taken by surprise. The settlements close to the location consist of a small number of families who have resided in the area for generations and some of whom still subsist from a combination of small-scale fisheries and other occupations (or are retired), and it constitutes a "second home" to many who are settled in the nearby larger towns of Honningsvåg and Alta.

In dialogues with Grieg Seafood and the Sámi Parliament, the residents raised unresolved private rights to the area, as well as consequences for coastal Sámi cultural survival in general. The issue of environmental influence from sea lice and pollution from the fish farm on coastal cod spawning grounds and adjacent wild salmon rivers (see Map 3, Figure 4.2) were brought up by the locals as well as by the Directorate for Fisheries and the county governor. Complaints and protests from the community led to a complicated licensing process, which ended in a decision by the Directorate of Fisheries to have the farm removed due to the lack of investigations into the implications of the establishment for Sámi culture and livelihoods (Eythórsson et al. 2019). Grieg Seafood argued that supporting the industry also supported a living and thriving coastal Sámi culture, referring to employment contributions, personal incomes and tax revenues to municipalities. Sámi employees at Grieg Seafood facilities also supported the development, making the argument that aquaculture and traditional fisheries can co-exist and that coastal Sámi communities should work together with the industry to ensure its sustainability. The fact that seven Sámi were employed at the facility, while only two to three inhabitants were registered as permanently settled in the area, were among the factors mentioned in a review of the decision by Rogaland County. The county was engaged to do the review, since Finnmark County owned an educational license at the site, and concluded with a decision to uphold the approval of the license (Rogaland fylkeskommune 2019). The location was also certified as "sustainable" under the Aquaculture Stewardship Council Salmon Standard (ASC 2020a) (as of May 26, 2019). The leader of a local Sámi resource center in the Porsanger area strongly criticized Grieg Seafood's use of the fact that Sámi were employed at the site as evidence of how aquaculture strengthens the material basis for Sámi culture in the area. In his perspective, the case was rather about the extent to which the site destroyed the livelihoods of Sámi fishers and impacted the natural basis for Sámi culture in the long term (Hansen 2019).

When the Vedbotn fish farm was established, the local inhabitants expressed disappointment in the Sámi Parliament's lack of ability to stop the fish farm licensing process. As the Sámi Parliament had previously been able to stop the installation of a fish farm in a comparable case (see Brattland and Eythórsson 2016), there were expectations that the Sámi people's own organ would speak on their behalf in the Vedbotn location licensing process. The Sámi Parliament had, however, not objected to the coastal zone plan when it was treated by the municipality board, because they did not have sufficient knowledge about the impacts of the planned location at the appropriate point in the planning process in order to intervene (Interview, June 25, 2018c). The Sámi Parliament made two formal complaints to the Directorate of Fisheries, upon which the site was unapproved in the first instance. After the Rogaland County's decision to continue the operation, the Sámi Parliament made the second complaint. The site continues to be controversial, while Grieg Seafood has continued to make efforts to improve relations with the local population in the area. This case is not exceptional as the Sámi Parliament is involved with around 30 to 40 aquaculture licensing processes every year, where some end up in conflicts with other sector authorities. As in the case of Vedbotn, this is often due to the lack of knowledge and criteria for what constitutes a serious enough impact on Sámi culture to weigh up for the economic benefits to the municipality at large.

Structural and discursive influences on Indigenous agency

The two cases we have narrated, although different in most aspects, both raise some commonalities that we identify as relevant factors influencing Indigenous agency as an aspect of equity in ocean governance. The structural factors that we see at the institutional level are first and foremost the lack of Indigenous rights recognition in the Norwegian case which constrains Sámi agency as compared to the Māori. Other factors are unequal access to resources, information and expertise which, in the case of Norway, leads to a lack of capacity to safeguard local actors. In the case of Aotearoa–New Zealand, the power dynamics inherent in the allocation of commercial space is more apparent as an outcome of the structural reality of the twenty-percent marine space allocation policy. Two discursive factors were identified as important in both cases at local and regional levels, namely the constraints set by discourses on development or protection of Indigenous rights, and perceptions of Sámi and Māori presence and connection to marine resources, which determine what is deemed as relevant knowledge in aquaculture development. In the NZ case, the issue of how Indigenous knowledge is dealt with in aquaculture licensing processes was more prominent than in the Norwegian case.

Legal constraints on Sámi and Māori agency

The lacking recognition of Sámi property rights and a consultation regime which recognizes the right of a larger group of Sámi inhabitants to be consulted on local development issues is a serious legal constraint on local Indigenous agency in Norway. This limitation is about to be addressed in the revised Sámi Act. In the Vedbotn case, the Sámi Parliament played a minor role, which illustrates how the lack of a legal structure makes communities dependent on the capacity of the Sámi Parliament to argue their case in licensing processes. This becomes especially evident compared to the strong role of the *iwi* in New Zealand and other Indigenous groups who have had their territorial rights recognized. Without the local inhabitants' complaints in Vedbotn, there is a possibility that Grieg Seafood had not noted any Indigenous presence in their application for sustainability certification to the Aquaculture Stewardship Council.

As both Grieg Seafood and Cermaq operate under the Norwegian coastal zone planning system, where local authorities have a key role in the licensing process, they do not have any obligations to consult with local communities directly. The ASC Salmon Standards, however, require a proactive consultation process, which is more than the Norwegian legislation requires. In both case study areas, salmon fish farms were certified by the ASC under the Salmon Standard. The ASC is an independent, non-profit organization that manages the certification and labeling program for responsible aquaculture. Consultation and dialogue between the industry and Indigenous peoples is a key requirement under criterion 7.2 for environmental sustainability and social responsibility, which pertains particularly to Indigenous peoples and local communities (Aquaculture Stewardship Council 2019). Thus, the ASC Salmon Standard criteria on local population and Indigenous peoples can be seen as an example of an enabling global structure which can enhance Indigenous agency at the local level given an awareness of Indigenous presence. Furthermore, the United Nations Declaration on the Rights of Indigenous Peoples (UNDRIP) and the ILO 169 oblige the state to consult with Indigenous peoples. These serve to enable the Sámi Parliament as an Indigenous political institution in Norway, but the lack of implementation of those rights at the local level leaves Sámi communities with less power to act than the Māori.

In Aotearoa–New Zealand, a structural constraint that impinges upon Indigenous agency is the power afforded to local councils under the Resource Management Act (RMA) of 1991. As mentioned previously, the RMA is the primary legislation which governs the establishment of marine farms. Any person, either Māori or Pakeha a white (non Māori) New Zealander, wishing to establish a marine farm must apply for and obtain a resource consent from the appropriate regional council or unitary authority. While Māori are allocated twenty percent of any new aquaculture space, as stipulated in the Māori Commercial Aquaculture Claims Settlement Amendment Act 2011, the ultimate decision-making power as to whether any new aquaculture spaces will be approved is held by the local council authority.

As the Sámi Parliament does not consult directly with fish farming companies and the Indigenous territorial marine rights of the Sámi have not been recognized, it is unclear to what extent companies operating in Norway are obliged to consult directly with local communities. As noted by Eythórsson et al. (2019), the municipal planning process in Norway both enables and constrains Sámi influence on fish farming, because objections by the Sámi Parliament are not a veto on a plan but have led to instances where fish farm location plans have been stopped (Brattland and Eythórsson 2016). The Sámi Parliament does not as a rule engage directly or consult with industry, as the Consultation Agreement (2005) is with the state authorities that regulate the activities of the industry. The Sámi Parliament can, however, based on the Planning and Building Act (2008), file formal objections to plans when they are sent out on hearings, although they have limited authority once the local municipality boards approve a plan.

Access to resources, information and expertise

Resource governance processes tend to favor the agency of those actors who have greater access to financial and technical resources and access to expertise and information, as in the case of Vedbotn. The Sámi Parliament's planning department has limited capacity to investigate each individual case, and there is no comprehensive database on Sámi marine usage and local Sámi interests that are potentially in conflict with salmon farming. In general, the assessment stage in the planning process is the best window of opportunity for Sámi fishers, communities and associations to influence decisions about aquaculture sites, also indicated in the Cermaq interview. However, Sámi marine interests span a large area to cover for the Sámi Parliament, reaching from the Lofoten archipelago and Tysfjord in the south to the Russian border in the east. A major constraint on the Sámi Parliament's agency is thus the lack of a comprehensive knowledge basis for their actions and lack of formalized contact with defined Sámi stakeholders. Another issue is timing. If the Sámi Parliament does not have access to the appropriate knowledge on Sámi interests early in the planning process, this will have repercussions for the duration of the case.

In Aotearoa–New Zealand there is a perception that salmon farming is environmentally harmful, and many local councils are opposed to new aquaculture developments in their regions. This perception of, and resistance to, salmon farming means that the agency of those Māori peoples who are not opposed to salmon farming and wish to participate in salmon farming opportunities through their twenty-percent allocation of aquaculture space is subject to the local council decision-making authority. The case of salmon farming in Marlborough Sounds was so contentious that an independent expert panel was appointed to assess the environmental and social impacts of any increased salmon farming in the region. They eventually found that the future for salmon farming lay in offshore ventures and was not appropriate for the Marlborough Sounds. Their finding flew in the face of the fact that new leases would be reallocating previous leases in more environmentally appropriate areas, which would have reduced the environmental impact even further, something NZKS were determined needed to be heard.

Yet despite this apparent structural constraint via the national legislative requirement that Māori are allocated twenty percent of all new aquaculture space, they can either partner in new developments or obtain cash settlements. Thus, the agency of those Māori who seek to participate in the economic opportunities presented by aquaculture developments may be seen to be privileged in this instance. While there are many scholars, such as the anthropologist Fiona McCormack (2017), who see this twenty-percent allocation as a neoliberal tool to encompass Māori in neoliberal opportunities, this is an entitlement for Indigenous peoples that is not legislated in Norway (although there are compensatory schemes such as an aquaculture fund to coastal municipalities). This legislative reality also ensures that Māori concerns will be addressed in decision-making processes as they are in fact a partner in any new aquaculture opportunity. Māori see the Treaty of Waitangi ensuring the acknowledgment of their partnership with the Crown in all future aquaculture developments. Their agency is therefore influenced by this legislated opportunity to gain materially from aquaculture development. Thus, there are legislative mechanisms at the local and the national level in NZ where Māori agency in aquaculture development has both constrained and enabled Māori.

Environmental protection and Indigenous development discourses

In Aotearoa–New Zealand there are several discourses that influence Māori agency in aquaculture developments. The discourse of salmon farming as intrinsically environmentally harmful influences local council decision-making processes, as many regional councils do not want this "problem" in their backyard. These local and regional councils are often if not always dominated by other (non-Māori) groups, marginalizing the Māori input into regional land and marine use decision-making processes. At the same time, for those Māori who seek to realize their mandated twenty-percent allocation by participating in salmon farming, this discursive trope can constrain their opportunities to participate in marine economic opportunities in their tribal regions. The reality that salmon farming practices in NZ are judged as the best in the world has little or no impact against this dominant discourse of salmon farming as environmentally harmful. Many of the interviewed Māori who were against the new aquaculture leases that would result from any relocation of existing farms in the Marlborough Sounds, cited the environmental unsustainability of salmon farming as incompatible with their tribal and cultural obligations and as their reason for refusing to approve any new leases.

Thus, this discourse of salmon farming as inherently environmentally negative constrains the agency of those Māori who wish to become economic players in the aquaculture industry and enables the agency of those Māori who oppose salmon farming. When distilling negative outcomes from aquaculture development, we identified a potential for community conflict. There are always opposing views in resource development opportunities, and those Māori who are not inclined to obtain material benefits from aquaculture developments, because of their perception that they are incompatible with their cultural and custodial obligations, are forced to defend their epistemological and cultural systems to the dominant cultural authorities and industry, when they oppose new aquaculture developments in their traditional marine spaces. Many of the same developments are true for Sámi areas. However, in comparative terms with Norway, Māori have a much stronger legislative basis for the incorporation of their epistemologies and cultural values in future salmon farming developments. This is not the case in Norway.

In the case of New Zealand, aquaculture is not as dominant in terms of marine development opportunities as in coastal Norway. Coastal Norway is characterized by conflicts between economic development and environmental protection, which are enhanced in areas affected by rural flight (Nilsen 2014). A salient discourse in the planning processes is thus the tension between the protection of the material basis for culture and developing an area.

At the local level, Sámi who are employed in the aquaculture industry will not have the same opportunity as the Māori to realize economic benefits from Indigenous ownership in the sector. In the Vedbotn case, those few Sámi who argued for development were met with opposition from local and Sámi politicians, and it is difficult for Sámi politicians with an environmental protectionist stance to gain votes in key constituencies with a pro-development discourse. The general lack of recognition of the importance of Sámi culture and livelihoods leads to the potential for community conflict between municipal pro-development govern-ments and coastal Sámi communities where fish farms are located, such as in the case of Vedbotn. The development discourse thus constrains Sámi political agency in the sense that it already makes it difficult for Sámi actors to have a meaningful role in aquaculture development other than being against it.

Discursive formation of relevant knowledge and claims

Discursive factors can determine what knowledge and what political claims are seen as possible or legitimate; they can determine who can make authoritative claims to hold such knowledge or make those political claims (Reimerson 2016). The most striking discursive influence in the Norwegian case is the lack of awareness among both public and private actors of what constitutes relevant Sámi interests in the coastal zone at the national, regional and local level. There is also no recognition of Sámi as development actors in the industry, which is recognized for the Māori. As the coastal zone planning system operates based on public knowledge, the lack of awareness is both a discursive and an institutional influence on agency.

This shapes what is perceived as relevant knowledge to be assessed in consulta-tions between the Sámi Parliament and the Norwegian government. The Sámi Parliament experiences a lack of government willingness to consult on aquaculture policy issues, and a lack of awareness of the impacts of aquaculture on the material basis for Sámi culture. The lack of awareness of what constitutes legitimate Sámi interests is also confirmed by the regional and municipal planning authorities (Eythórsson et al. 2019). At the municipality level, we also need to ask to what extent local politicians and local branches of the aquaculture companies recognize or silence awareness of Sámi presence.

The privileging of scientific knowledge over Māori knowledge affects Māori agency in aquaculture development in Aotearoa–New Zealand. In this NZ case study, Māori representatives drew on their traditional knowledge to oppose NZKS' application for the aquaculture leases. Their role, duties and rights under the prin-ciple of *Kaitiakitanga*—to act as guardians of the environment—came to the fore in this case study and also underpinned the concerns and submissions made to the panel by the majority of *iwi* in the region, including Ngāti Kuia, Ngāti Koata, Ngāti Toa and Ngāti Apa, which resisted the expansion and relocation of salmon farming in their *rohe*.

Thus, in the NZ case, a discourse that is often evident in many resource decision-making processes involving Indigenous peoples, the discourse of scientific knowledge as somehow superior to Māori knowledge was evident in the decision-making processes for salmon farming in the Marlborough Sounds. That the CEO of NZKS questioned the integrity of this traditional knowledge is also familiar in many resource decision-making processes on Indigenous lands and waters. This discourse privileges one knowledge system over another, the scientific over the Indigenous knowledge system, and ultimately favors the agency of those who have greater access to both scientific knowledge and resources.

Conclusion

We started this chapter by asking how Indigenous agencies are influenced by structural and discursive factors in the two different contexts of northern Norway and Aotearoa–New Zealand, and how Indigenous power to act for equity can be strengthened.

Starting with the legal factors, the process of consultation and participation in aquaculture licensing differs greatly between Norway and NZ. In essence, the legal recognition of local Indigenous governance institutions such as the *iwi* in New Zealand clearly enhances Indigenous agency to enter into consultations with and participate in aquaculture development initiatives. In NZ, the local *iwi* are the natural and relevant consultation partners for the industry and state agencies alike, while the lack of a right to consult for local communities makes this unclear in the Norwegian case. Our case studies also illuminate the importance of local agency in aquaculture development. The lack of state recognition of Sámi presence and interests in the coastal zone seriously hinders Sámi agency in aquaculture development both locally and more broadly. Even though transnational aquaculture companies do not have direct human rights obligations under international law, they may play important roles as drivers of equity through certification and compensatory schemes. Greater attention to both human rights and the environment by legislators, combined with improved corporate reporting and increased transparency in global supply chains, is incentivizing corporations to operate responsibly (Folke et al. 2019).

The principles set by global standards such as the UN Global Compact and international expectations of private businesses to adhere to the duty to consult and have agreements with Indigenous local communities influence Indigenous agency in cases where these are implemented successfully. Examples include the ASC Salmon Standard and the recognition of Māori rights in the New Zealand governance system, which serve to strengthen Indigenous agency. However, as noted above, this can constrain the agency of pro-development Māori individuals and organizations. Thus, there should be a greater recognition at the local level (in the Resource Management Act) of Māori rights to engage with aquaculture developments if they do wish to have economic opportunities. The role of the Sámi

Parliament as a "watchdog" on all municipal plans holds the potential to strengthen Indigenous local agency in Norway, but as noted in the interview with the Sámi Parliament, there is a clear need to recognize the right of local communities to consult on issues that affect Sámi interests and livelihoods. Structural constraints such as access to resources, information and expertise play a bigger role for small, remote, rural Indigenous communities than for larger Indigenous communities in more populated areas. Despite drawbacks, the Māori experiences with the twenty-percent share arrangement could serve as inspiration for the Norwegian discussion on how redistribution of the aquaculture fund to municipalities with aquaculture activities could consolidate Sámi agency.

A possible factor that could enhance Indigenous agency is the method by which farms are approved as sustainable. Social responsibility makes up only a minority of the total criteria on which farms are audited as sustainable by the ASC. Instead of weighting social responsibility criteria as a part of the total, an alternative approach could be to give equal weight to both social and environmental criteria to achieve social equity. Including social equity in sustainability assessments could strengthen Indigenous agency and enhance the sustainability of the ocean economy globally (Österblom et al. 2020). The broader governance situation also needs to be considered, such as the lack of consultations with the Sámi Parliament on national aquaculture policies in Norway. In the case of New Zealand, there were no complaints in the particular ASC licensing process for the Clay Point farm, but numerous complaints on the planning of aquaculture in Marlborough Sounds in general (ASC 2020b). This points to a need for broader assessments of Indigenous agency not only at the farm site level but also including how social responsibility is dealt with at the related governance levels associated with the farm licensing process.

In both cases, development discourses tend to constrain rather than enable Indigenous agency. The arguments of those who are against development are trapped in a structure that makes it difficult to compete with the weight of economic development against the "light" weight of social responsibility and marine resource rights. Following Howlett's (2010) argument that discursive formations constrain Indigenous agency in certain ways, the discursive framing of development as exclusively non-Indigenous may constrain Indigenous peoples' agency in ways that make it more difficult to be both Indigenous and development-oriented at the same time. While the natural role of the Māori marine resource rights acts to enable knowledge and recognition of Māori as players in the aquaculture sector in New Zealand, the lack of knowledge and outright ignorance at many levels in the Norwegian case is a serious constraint on Sámi agency both in the local communities and more broadly. The Māori agency, however, extends only to a certain level, as Indigenous knowledge is still disputed as a valid knowledge basis in aquaculture licensing processes. Greater acceptance and recognition of Indigenous knowledge as a legitimate system of knowledge would enhance Indigenous agency in both cases.

Summing up, our discussion leads us to conclude that Indigenous agencies are more constrained than enabled by structural and discursive factors impinging upon aquaculture development in Indigenous marine territories. This serves to heighten the ongoing critique of unsustainable neoliberal forms of development. What could be done to strengthen Indigenous agency and increase the sustainability of industrial development projects? We do not support capacity and agency by telling Indigenous youngsters that they cannot take on that work in a mine or on a fish farm because it does not support Indigenous life projects (Blaser et al. 2004). Neither do we support Indigenous agency by not providing a realistic alternative. For those who are pro development, few resources are available for discussing development on Indigenous terms, or even for discussing how aquaculture could contribute to the well-being of not only Indigenous communities, but of nature and the planet. As ocean-based developments particularly affect coastal communities, the capacity of those communities to negotiate benefits from ocean development may be key to their existence. The ability to benefit from ocean resources, however, requires equality of opportunity, which again is provided by human, social, financial and physical assets (Sen 1992; Nussbaum 2011; Bennett et al. 2018).

We propose that states and private actors take a more proactive role in their dealings with Indigenous–industry relations and incorporate Indigenous development agendas into their practices. This entails going beyond defining the outcome of development in terms of economic benefits such as employment or measuring Indigenous agency as the power to stop development only. Rather, we encourage Indigenous participation at the stage where policies and priorities are formed. What would a salmon farming operation based on Indigenous values and priorities look like? What would be the priorities formulated by Indigenous actors in determining the sustainability of multinational companies? Allowing space for having questions such as these discussed and included, if not answered, would in our opinion enhance Indigenous agency as the power to act and participate meaningfully in aquaculture development. As this chapter has illustrated, the Māori are already a step ahead in participating in development at the national level, while the right to be consulted is still in the making for Sámi communities.

References

Aquaculture Act (2006). *Lov om akvakultur (akvalkulturloven)*. Norway: Ministry of Commerce and Fisheries.

Åhrén, M. (2016) *Indigenous peoples' status in the international legal system*. Oxford: Oxford University Press.

Allen, S., Bankes, N. and Ravna, Ø. (eds.) (2019) *The rights of Indigenous peoples in marine areas*. London: Hart Publishing.

Aquaculture Stewardship Council, ASC (2019) *ASC salmon standard: version 1.3* [Online]. Available at: https://www.asc-aqua.org/wp-content/uploads/2019/12/ASC-Salmon-Standard_v1.3_Final.pdf (Accessed January 1, 2020)

ASC (2020a) Grieg Seafood Finnmark AS: ASC-DNV-288534. *Vedbotn Farm sustainability license* [Online]. Available at: https://www.asc-aqua.org/find-a-farm/ASC00901/ (Accessed December 21, 2020)

ASC (2020b) New Zealand King Salmon ASC-SGS-F-025. *Clay Point Farm sustainability license* [Online]. Available at: https://www.asc-aqua.org/find-a-farm/ASC01359/ (Accessed December 21, 2020)

Bargh, B. (2016) *The struggle for Māori fishing rights*. Wellington: Huia Publishers.

Bargh, M. (2018) "Māori political and economic recognition in a diverse economy" in D. Howard-Wagner, M. Bargh and I. Altamirano-Jiménez (eds.), *The neoliberal state, recognition and Indigenous rights: new paternalism to new imaginings.* Canberra: Centre for Aboriginal Economic Policy Research, The Australian National University, pp. 293–307.

Bartlett, L. and Vavrus, F. (2016) *Rethinking case study research: a comparative approach* [Online]. Taylor & Francis eBooks. Available at: https://doi.org/10.4324/9781315674889

Bell, A. (2018) "A flawed treaty partner: the New Zealand state, local government and the politics of recognition" in D. Howard-Wagner, M. Bargh and I. Altamirano-Jiménez (eds.), *The neoliberal state, recognition and Indigenous rights: new paternalism to new imaginings.* Canberra: Centre for Aboriginal Economic Policy Research, The Australian National University, pp. 77–91.

Bennett, N.J., Kaplan-Hallam, M., Augustine, G., Ban, N., Belhabib, D., Brueckner-Irwin, I., Charles, A., Couture, J., Eger, S., Fanning, L., Foley, P., Goodfellow, A.M., Greba, L., Gregr, E., Hall, R., Harper, S., Maloney, B., McIsaac, J., Wanli, O, Pinkerton, E., Porter, D., Sparrow, R., Stephenson, R., Stocks, A., Rashid Sumaila, U., Sutcliffe, T. and Bailey, M. (2018) "Coastal and Indigenous community access to marine resources and the ocean: a policy imperative for Canada," *Marine Policy,* 87, pp. 186–193.

Blaser, M., Feit, H.A. and McRae, G. (eds.) (2004) *In the way of development: Indigenous peoples, life projects, and globalization.* London and New York: Zed Books.

Brattland, C. (2013) "Proving fishers right: effects of the integration of experience-based knowledge in ecosystem-based management," *Acta Borealia,* 30(1), pp. 39–59.

Brattland, C. and Eythórsson, E. (2016) "Fiskesløyfa: Spildrafiskernes driftsformer og oppdrettsaktiviteten" [The fishing loop. Spildra fishers' practices and the aquaculture industry], *Ottar,* 4, pp. 23–33. Tromsø: Tromsø Museum.

Brattland, C., and Nilsen, S. (2011). "Reclaiming indigenous seascapes. Sami place names in Norwegian sea charts," *Journal of Polar Geography,* 34(4), pp. 275–297.

Broderstad, E.G. (2014) "Implementing Indigenous self-determination: the case of the Sámi in Norway" in Woons, M. (ed.) *Restoring Indigenous self-determination: theoretical and practical approaches.* Bristol: E-International Relations Publishing, pp. 1–6.

Broderstad, E.G. and Eythórsson, E. (2014) "Resilient communities? Collapse and recovery of a social-ecological system in Arctic Norway", *Ecology and Society,* 19(3), pp. 1–10.

Consultation Agreement (2005) *Avtale om prosedyrer for konsultasjoner mellom statlige myndigheter og Sametinget* [Online]. Available at: https://www.regjeringen.no/en/topics/indigenous-peoples-and-minorities/Sami-people/midtspalte/PROCEDURES-FOR-CONSULTATIONS-BETWEEN-STA/id450743/ (Accessed August 19, 2020)

Dahl, J. (2012) *The Indigenous space and marginalized peoples in the United Nations.* New York: Palgrave Macmillan.

Draft resolution and bill 2020–2021. Prop. 86 L (2020–2021) *Proposisjon til Stortinget, Endringer i sameloven mv. (konsultasjoner)* [draft resolution and bill, changes in the Sámi Act (consultations)] [online]. Available at: https://www.regjeringen.no/no/dokumenter/prop.-86-l-20202021/id2835131/ (Accessed February 27, 2021).

Evjen, B. (2014) "Åpning for sjøsamiske rettigheter" in Kolle, N. and Christensen, P. (eds.) *Norges fiskeri- og kysthistorie bind IV* [Opening for coastal Sami rights. Norwegian fisheries and coastal history, part IV]. Bergen: Fagbokforlaget, pp.251–279.

Eythórsson, E., Schreiber, D., Brattland, C. and Broderstad, E.G. (2019) "Governance of marine space: interactions between the salmon aquaculture industry and Indigenous peoples in Norway and Canada" in S. Allen, N. Bankes and Ø. Ravna (eds.) *The rights of Indigenous peoples in marine areas.* London: Hart Publishing, pp. 353– 374.

Finnmark Act (2005) Lov 17. juni 2005 nr. 85 om rettsforhold og forvaltning av grunn og naturressurser i Finnmark fylke A - Act of 17 June 2005 No 85 Relating to Legal Relations and Management of Land and Resources in the County of Finnmark.

Folke, C., Österblom, H., Jouffray, J.B., Lambin, E.F., Adger, W.N., Scheffer, M., Crona, B.I., Nyström, M., Levin, S.A., Carpenter, S.R. and Anderies, J.M. (2019) "Transnational corporations and the challenge of biosphere stewardship," *Nature ecology & evolution*, 3(10), pp.1396–1403.

Hansen, T. (2019) "Grieg Seafood, Rogaland fylkeskommune og sjøsamisk etnisitet" [Grieg Seafood, Rogaland County and coastal Sámi ethnicity] [Online], *Dagsavisen*, November 4. Available at: https://www.dagsavisen.no/debatt/2019/04/11/grieg-seafood-rogaland-fylkeskommune-og-sjosamisk-etnisitet/ (Accessed January 1, 2020)

Hersoug, B. (2002) *Unfinished business: New Zealand's experience with rights-based fisheries management*. Delft: Eburon.

Hersoug, B. (2005) *Closing the commons: Norwegian fisheries from open access to private property*. Delft: Eburon.

Hersoug, B. (2018) "'After all these years'—New Zealand's quota management system at the crossroads," *Marine Policy*, 92, pp. 101–110.

Hersoug, B. and Johnsen, J.P. (2012) *Kampen om plass på kysten: interesser og utviklingstrekk i kystsoneplanleggingen* [The fight for coastal space: interests and development trends in coastal zone planning]. Oslo: Universitetsforlaget.

Hovland, E. (2014) "Havbruksnæringen i krise 1989–1991 [Aquaculture industry in crisis 1989–1991]" in E. Hovland, D. Møller, A. Halland, N. Kolle, B. Hersoug, and G. Nævdal (eds.) *Over den leiken ville han rå. Norsk havbruksnærings historie. Bind V.* [The game for which he wanted to master: the history of Norwegian fish farming]. Bergen: Fagbokforlaget, pp. 215–249.

Howlett, C. (2010) "Indigenous Agency and Mineral Development: A Cautionary Note," *Studies in Political Economy*, 85, pp. 99–123

Inns, J. (2018a) "Marlborough Salmon Farm relocation proposal," *Aquaculture, A82 March/April*, pp. 6–7.

Inns, J. (2018b) Personal communication, *senior legal representative of Ocean Law. Nelson, New Zealand*, April 12.

Interview (2018a) Bergan, L., Cermaq office, Oslo, Norway, May 25.

Interview (2018b) Elkington, G., phone interview, New Zealand April 2.

Interview (2018c) Muotka S.K., Fram Centre, Tromsø, Norway June 25.

Interview (2018d) Rosewarne, G. and Gillard, M., Nelson, New Zealand, April 17.

Interview (2018e) Ruru, H., Nelson, New Zealand, April 16.

Jentoft, S. and Eide, A. (2011) *Poverty mosaics: realities and prospects in small-scale fisheries.* Amsterdam: Springer Science & Business Media.

Josefsen, E. (2014) *Selvbestemmelse og samstyring: en studie av Sametingets plass i politiske prosesser i Norge* [Self-determination and co-governance: a study of the role of the Sámi Parliament in political processes in Norway]. Doctoral dissertation. UiT Norges arktiske universitet: Fakultet for humaniora, samfunnsvitenskap og lærerutdanning.

Marine Resource Act (2009). *Lov om forvaltning av viltlevende marine ressurser (havressurslova).* Norway: Ministry of Commerce and Fisheries.

McAnnulla, S. (2002) "Structure and Agency," in D. Marsh and G. Stoker (eds.) *Theory and Methods in Political Science* 2nd ed. Hampshire, UK: Palgrave Macmillan.

McCormack, F. (2017) "Sustainability in New Zealand's quota management system: a convenient story," *Marine Policy*, 80, pp. 35–46.

McGinnis, M.V. and Collins, M. (2013) "A race for marine space: science, values, and aquaculture planning in New Zealand," *Coastal Management*, 41(5), pp. 401–419.

McPhee E. (2017) "Marlborough iwi plan to raise sea guardian in fight to stop salmon farm relocation," *Stuff*, April 16 [Online]. Available at: https://www.stuff.co.nz/business/91547386/Marlborough-iwi-plan-to-raise-sea-guardian-in-fight-to-stop-salmon-farm-relocation (Accessed: May 27, 2019).

Meihana, P. (2017) *Submission to the Marlborough Salmon Farm relocation advisory panel public hearing (MSFRAPPH). Transcript of proceedings.* Blenheim: Marlborough Convention Centre.

Minister for Primary Industries (MPI) (2017) *Report and recommendations of the Marlborough Salmon Farm relocation advisory panel.* Wellington, New Zealand.

Mulrennan, M. and Scott, C. (2000) "Mare nullius: Indigenous rights in saltwater environments," *Development and Change*, 31(3), pp. 681–708.

Nilsen, R. (2014) "Rural modernisation as national development: the Norwegian case 1900–1950," *Norsk Geografisk Tidsskrift–Norwegian Journal of Geography*, 68(1), pp. 50–58.

NIM, (2016) Norges nasjonale institusjon for menneskerettigheter (*Temarapport 2016. Sjøsamenes rett til sjøfiske* [Thematic report 2016. The right of coastal Sámi to fish in Norway's offshore waters] [Online]. Oslo: NIM. Available at: https://www.nhri.no/2017/temarapport-2016-sjosamenes-rett-til-sjofiske/ (Accessed December 18, 2020)

Nussbaum, M.C. (2011) *Creating capabilities: the human development approach.* Cambridge, MA: Harvard University Press.

O'Faircheallaigh, C. (2016) *Negotiations in the Indigenous world: Aboriginal peoples and the extractive industry in Australia and Canada.* New York: Routledge

Österblom, H., Wabnitz, C.C.C., Tladi, D., Allison, E.H., Arnaud-Haond, S., Bebbington, J., Bennett, N., Blasiak, R., Boonstra, W., Choudhury, A., Cisneros-Montemayor, A., Daw, T., Fabinyi, M., Franz, N., Harden-Davies, H., Kleiber, D., Lopes, P., McDougall, C., Resosudarmo, B.P. and Selim, S.A. (2020) *Towards ocean equity* [Online]. Washington, DC: World Resources Institute. Available at: www.oceanpanel.org/how-distribute-benefits-ocean-equitably (Accessed January 1, 2020)

Planning and Building Act 2008. *Lov om planlegging og byggesaksbehandling* (Plan- og bygningsloven), LOV-2008-06-27-71. English translation available at: https://www.regjeringen.no/en/dokumenter/planning-building-act/id570450/ (Accessed June 20, 2020)

Reimerson, E. (2016) "Sami space for agency in the management of the Laponia World Heritage site," *Local Environment*, 21(7), pp. 808–826.

Rogaland fylkeskommune (2019) *Tillatelse – Grieg Seafood Finnmark AS – org.nr. 908361306 – akvakultur av matfisk av laks, ørret og regnbueørret – lokalitet nr. 10652 – Vedbotn – Nordkapp kommune. Fylkesrådmannen, næringsavdelingen* [Approval – Grieg Seafood Finnmark AS – org.nr. 908361306 – aquaculture of salmon, trout and rainbow trout for human consumption – location nr. 10652 – Vedbotn – Nordkapp municipality]. Letter from Rogaland County to Grieg Seafood Finnmark AS, March 28.

Ruru, J. (2010) "Claiming native title in the foreshore and seabed" in L.A. Knafla and H. Westra (eds.) *Aboriginal title and Indigenous peoples: Canada, Australia, and New Zealand.* Vancouver: UBC Press, pp. 185–201.

Sandersen, H. (2017) "The role of aquaculture in local communities, the model of Musken," in Moriarty, C.. *TriArc Tysfjord visit summary.* Tromsø: TriArc project, Centre for Sami Studies, pp. 7–8.

Sen, A. (1992) *Inequality re-examined.* Oxford: Oxford University Press.

Smith, C. (2014) "Fisheries in coastal Sami areas: geopolitical concerns?" *Arctic Review on Law and Politics*, 5(1), pp. 4–10.

Smith, R. (2017) *Submission to the Marlborough Salmon Farm relocation advisory panel public hearing (MSFRAPPH). Transcript of proceedings.* Blenheim: Marlborough Convention Centre.

Statistics New Zealand (2021) *Marlborough region* [Online]. Available at: https://www.stats.
govt.nz/place_summaries Place Summaries | Marlborough Region | Stats NZ (Accessed
January 1, 2020)

Statistics Norway (2021) *Kommunefakta* [Municipal facts] [Online]. Available at: www.ssb.
no/kommunefakta (Accessed January 1, 2020)

Tennberg, M. (2010) "Indigenous peoples as international political actors: a summary," *Polar
Record*, 46(3), pp. 264–270.

Trebeck, K. (2007) "Tools for the Disempowered? Indigenous Leverage Over Mining
Companies," *Australian Journal of Political Science*, 42(4), pp. 541–62.

Yin, R.K. (2012) *Applications of case study research*. Thousand Oaks, CA: SAGE.

Young, N., Brattland, C., Digiovanni, C., Hersoug, B., Johnsen, J.P., Karlsen, K.M., Kvalvik, I.,
Olofsson, E., Simonsen, K., Solås, A.-M., and H. Thorarensen (2019) "Limitations to
growth: Social-ecological challenges to aquaculture development in five wealthy nations,"
Marine Policy, 104, pp. 216–224. Available at: https://doi.org/10.1016/j.marpol.2019.
02.022

5 Indigenous agency through normative contestation

Defining the scope of free, prior and informed consent in the Russian North

Marina Peeters Goloviznina

Introduction

In March 2014, Almazy Anabara, a subdivision of ALROSA, the world leader in diamond mining, obtained a license in the Olenek Evenks county, the Republic of Sakha (Yakutia) RS (Ya). Neither the local district (*ulus*) administration nor the residents of the village of Zhilinda were informed about the planned mining. Zhilinda, in which the vast majority of the population are Evenks, has the status of a *territory of traditional nature use* (TTNU), which grants its Indigenous residents a right similar to free, prior and informed consent (FPIC) or *svobodnoye, predvaritel'noye i osoznannoye soglasiye*. The community gave the company its consent only for three of the four proposed mining sites. The locals protested mining on the Malaya Kuonapka River, a sacred place for Evenks and the only source of drinking water and fish. The *ulus* administration summoned the federal Agency for Subsoil Use to arbitration and demanded them to cancel the results of the auctions at Malaya Kuonapka for violating Indigenous peoples' rights under the TTNU law for FPIC. Despite the public outcry, the arbitration found no violation of the Evenks' right to FPIC.

Free, prior and informed consent was outlined in the Indigenous and Tribal Peoples Convention No. 169 (ILO Convention 169) and fully introduced by the UN Declaration on the Rights of Indigenous Peoples (UNDRIP) as a specific Indigenous peoples' right to self-determine through meaningful consultation on how a project may affect them or their territories. Over the past decades, FPIC has become a global normative umbrella principle with growing yet contested recognition among governments and corporations to secure Indigenous peoples' rights in an extractive context. Free, prior and informed consent is still an evolving international norm: its normative status is not clear enough, and its procedural implementation is controversial (Heinämäki, 2020, p. 335).

While the Russian Indigenous representatives and diplomats took an active role in the work on the UNDRIP, Russia has refrained from endorsing the declaration and has not ratified ILO 169. The above legal case history from the RS (Ya) FPIC shows that it has found its way into deliberations on the Russian ground. It also demonstrates how FPIC performs in the RS (Ya) and how Indigenous peoples strive to use this international tool to defend their rights regarding local mining.

Scholars have recently begun to delve deeper into studying international (soft) regulations in the Russian extraction context, recognizing their growing importance

DOI: 10.4324/9781003131274-5

and use over the past decade (Novikova and Wilson, 2017). Some studies show how engagement with global markets (supply chains, funding) and adherence to international corporate regulations have changed companies' conduct toward Indigenous peoples at the local level (Stammler and Wilson, 2006; Tulaeva et al., 2019). Others have taken a bottom-up approach to examine how the development of international regulations, globalization and the growth of Indigenous activism and information technologies have affected Indigenous peoples' participation in and control over resource development (Tysiachniouk et al., 2018).

Research findings on these issues are mixed. Some scholars argue that international Indigenous peoples' rights and ethical guidelines for industry performance are not well known among Indigenous stakeholders (Stammler et al., 2017). Others highlight cases when Indigenous peoples' organizations (IPOs) have voiced local injustices in the language of international Indigenous rights and even managed to "catch the moment" to improve their position (Peeters Goloviznina, 2019). Indeed, the debate on how to study Indigenous actors' perceptions on such complex issues as FPIC needs even greater scholarly attention (see the discussion on human security, Hoogensen Gjørv and Goloviznina, 2012, pp. 2–3).

Free, prior and informed consent is the latest addition to the Russian debate. As scholarship on the concept in the Russian context remains limited, there is much to be explored on the history of FPIC institutionalization and its encounter with domestic IPOs. How do Russian IPOs perceive and interpret FPIC? What is their experience of it and its implementation on the ground? More importantly, can the IPOs use the regulative power of the FPIC to ensure greater participation and control by their constituents over their homelands' developments?

This study contributes to the growing branch of scholarship examining encounters with FPIC from the perspective of the most numerous and diverse types of grassroots IPOs in contemporary Russia – *obshchiny* (often translated literally as nomadic clan communities). The study takes a bifocal research perspective, both normative and empirical, to explore the role of *obshchiny* in enabling the right of their constituents to FPIC in extractive projects in the Russian North. The Russian Federal Law No 104-FZ defines *obshchiny* as "a kinship-, family- or community-based organization of Indigenous peoples, formed to protect their traditional territories, traditional ways of life, culture, rights, and legal interests" (Russian Federation, 2000). In addition to their large number (1,597 *obshchiny* registered in Russia), the choice of *obshchiny* also has another analytical reasoning (Russian Federation, 2020). Given the specifics of the Russian approach to recognizing Indigenous peoples' territorial rights, *obshchiny* are the only legal entity through which the state recognizes Indigenous peoples' collective rights to land and use of resources (Kryazhkov, 2015).

Over the last decades, scholars have produced two different, albeit interrelated, narratives in studying the *obshchiny*. One concerns the historical (imperialistic) legacies and structures (institutions and power) of Russian Indigenous politics, limiting the possibilities of *obshchiny* to ensure their constituents' rights to land, autonomy and self-determination. The other narrative is about how the Indigenous organizations' lack capacity to take advantage of new opportunities (globalization, digital revolution) to realize the aspiration for economic, cultural and social advancements.

Subscribing to both narratives, I argue that they belong to just one side of the story about IPOs from *above*, a perspective of those with dominant status in power relations. To complement this mainstream yet one-way approach, I suggest rethinking the agency of IPOs from another angle, from *below*. The actors-based perspective spotlights the tactical, instrumental and localized practices the IPOs use to contest the normative roots that regulate their relations with the more powerful and resourceful counterparts. Incorporating these organizations' voices into the mainstream top-down debate will make more visible the processes of normative and social change they initiate and engage in from the bottom up. This advances our understanding of IPOs' agency in the context of the rights-flawed Russian state.

The study's empirical part is designed as a case study of the relationship between a family-based Evens *obshchina* and a gold mining company in the Republic of Sakha (Yakutia) in 2015–2019. Zooming into the practice of normative contestation around FPIC, I explore how the *obshchina*, contesting the company's visions on FPIC, was able to secure an advantageous interpretation of it; and how, under the prevailing unfavorable circumstances, the *obshchina* was able to maximize its benefits and interests. The choice of the RS (Ya) for the study has methodological reasons. Scholars demonstrate a consensus, acknowledging the republic as an "outstanding" case due to its Indigenous legislation's progressiveness and advanced law enforcement mechanisms to regulate "Indigenous–industries" relations. The study contributes to the scholarship, highlighting the institutional mechanisms behind the "advanced," rights-based approach to Indigenous politics.

The article consists of six sections. Following the introduction, the second part outlines a theoretical framework, sketching the ideas on agency and norms in a normative contestation analysis. The third part describes the methodology and methods used. The next sections examine the specifics of FPIC in the Russian legal framework and discuss the case study findings from the RS (Ya). The final part ends with the conclusions.

Agency and norms through practice of normative contestation

The ontological ground of FPIC lies in the right of Indigenous peoples to self-determination: through their representative organizations, Indigenous peoples have the right to express their views and decide what happens on their lands, exerting control and governing these developmental activities (Heinämäki, 2020, p. 345). The normative foundation of the FPIC process is based on the ideas of participatory citizenship and democratic governance (Hajer and Wagenaar, 2003; Kooiman et al., 2005). My analysis joins this stream of scholarship examining how Indigenous actors challenge the existing norms to bring about social and political change in governance. By centering attention on the Indigenous agency's encounter with the norm of FPIC, I apply a norm contestation analysis (NCA) (Wiener, 2014; Jose, 2018).

Norm contestation analysis originates from international relations (IR) norm scholarship that concerns norms and norm-related behavior across global–local scales (Wiener, 2014). This analysis considers contestation as a "social practice that discursively expresses disapproval of norms and entails objection to them"

(Wiener, 2014, p. 30). It acknowledges the diversity of norms and their crucial role in regulating actors' social behavior (states, organizations, individuals). While mainstream IR scholarship focuses on studying norms at the international level, other scholars contribute with insights from normative contestation behavior at the micro-scales of a global society (Deitelhoff and Zimmermann, 2018).

Instead of viewing the norms as stable, the approach emphasizes their dual nature (quality), which implies that they are both structuring (stable) and socially constructed (flexible) (Wiener, 2014, pp. 19–24). Understanding norms as dynamic constructs of dual quality foregrounds the relationship between norms and agency. Norms never remain valid by themselves; they need constant affirmation by the actors through their practice. Hence, the actors can always (re-)produce the dominant meaning of the norms or contest it.

Agency manifests the norm-generating power of actors, which derives from and is exercised through actors' asymmetrical relations as power-holders engaged in a normative contestation (Wiener, 2014, p. 9). Cultural contexts and institutional arenas, varying significantly, play a critical role in enabling the actors' agency to contest the existing norms. The *presence of institutional mechanisms* that facilitate the participation of actors (stakeholders) in contestation processes and the *access* of actors to them largely determine the actors' ability to exercise their norms-generating power. The power of those with limited or without institutional access to the normative contestations sites and mechanisms remains negligible and restricted (Wiener, 2017, p. 12).

Scholars consider NCA particularly useful for examining human behavior related to ambiguous norms (both social and legal) and interactions they have caused (Jose, 2018, p. 34). When international norms touch the ground in a given context, they generate multiple interpretations of their content, prescriptions (what the norm enables and prohibits) and their parameters (the situations in which the norm applies) (Jose, 2018, p. 5). Relatively, they encourage and enforce the actors, as norm-followers, to operationalize the meaning of these norms and define appropriate, norm-compliant behavior.

International Indigenous rights fall into the category of norms whose ambiguity plagues their conceptualization and challenges their practical application. What is FPIC, then? How should it be performed on the ground, by whom and under what conditions? The vague articulation of FPIC as a normative concept within international documents makes it an ideal target for contestation by Indigenous actors and extractives. With different backgrounds, driven by diverse (even adverse) interests, these actors have a conflicting interpretation of FPIC. While studies show that current FPIC practice is replete with positive and negative examples, the scholars also highlight its potential for negotiating mutually beneficial agreements (Rombouts, 2014, p. 23).

Research methodology and methods

This study was informed by data collected in fieldwork and desk research and primarily applied qualitative techniques, including semi-structured interviews, participatory observation and document analysis. In total, twenty-two interviews were conducted to clarify the informants' perceptions of FPIC and related issues (consultation, consent, benefits-sharing) in the Republic of Sakha. A large part of the interviews was conducted during the fieldwork in two settings: in Yakutsk (February–March 2019) and the *obshchina* winter camp along the Verkhoyansk Range (March 2019). Among my informants were the *obshchina* members, representatives of the republican authorities, the Ombudsman for Indigenous Peoples' Rights (OIPR), regional branches of Indigenous public organizations, including the Association of Indigenous Peoples of the North (AIPON), the World Reindeer Herders Association (WRH), the Union of the Nomadic Obshchiny (UNO) and academia. The names of many informants were anonymized to protect their identity. Most of the interviews were conducted in Russian, recorded and transcribed as text documents.

The secondary data for analysis is a corpus of official documents on Indigenous issues, including the relevant federal and RS (Ya) legislation, policy papers and the reports of the OIPR (2014–2019) (OIPR, 2020). The open-access data on Polymetal's social and Indigenous policy was obtained through the company's website (Polymetal International plc, 2020). These data have also been coded, categorized and analyzed using a mix of interpretative analysis techniques.

The challenge of FPIC in the Russian context

Although Russia has not endorsed the UN Declaration on the Rights of Indigenous Peoples (UNDRIP), it has reaffirmed its commitment to FPIC on numerous international platforms (OHCHR, 2018). Russian officials have always emphasized that the FPIC has to be interpreted through the normative lens of national legislation. The status and rights of Indigenous peoples are enshrined in the Constitution of Russia (1993) and three federal Indigenous laws, namely "On the Guarantees of the Rights" (1999), "On Organization of *Obshchiny*" (2000) and "On Territories of Traditional Nature Use" (2001) (Russian Federation, 1992, 1993, 1999, 2000, 2001). This legal framework incorporates Russia's approach to recognizing "Indigenous peoples" and their land rights.

At the core of Russia's approach to recognizing indigeneity lies the concept of *korrennye malochislennye narody Severa, Sibiri i Dalnego Vostoka*, KMNS (small-numbered peoples of the North, Siberia and the Far East). The law defines KMNS as

> peoples living in the territories of their ancestors' traditional settlements, preserving the traditional way of life and economic activities, numbering fewer than 50,000 persons, and recognizing themselves as independent ethnic communities.
>
> (Russian Federation, 1999)

Forty ethnic groups have KMNS status and represented 0.2 percent of the country's population at the last census (Russian Federation, 2010).

Russia's approach to recognizing KMNS land rights also differs from other Arctic states (Fondahl et al., 2020). They live and maintain their economies in a gigantic area rich in natural resources. Much of the land is public property, as the territory is vital for Russian national security and its resources-based economy. The state does not recognize the inherent rights to ancestral lands of small-numbered peoples of the North, Siberia and the Far East, but only their usufruct rights to land tenure (where the title remains with the state).

Russia has no particular law on FPIC. The legislator grants the scope of FPIC-related rights to *obshchiny*, recognizing them as the only rights-holders of the KMNS collective rights (Kryazhkov, 2015). The modern institutional history of the *obshchiny* has its origins in post-Soviet Russia (Fondahl et al., 2001; Gray, 2001; Novikova, 2001; Stammler, 2005; Sirina, 2010). With the crash of the Soviet command economy and the system of state farms (*sovkhozes*), the land from state farms (but not property rights) was transferred to the *obshchiny*. In the 1990s, the *obshchiny* registered their legal entities as various commercial agricultural organizations (Sirina, 2010; Stammler, 2005).

The Presidential Edict of 1992 issued two directives of a revolutionary character (Russian Federation, 1992). The edict called on the regional governments to transfer reindeer pastures, hunting grounds and fishing areas used by KMNS to their *obshchiny* for "life-time possession, free-of-charge use" (Russian Federation, 1992). The edict also called on the authorities to define the TTNU and declare their indefeasible status for any extractive activities. Since then, the institutional linkup between *obshchiny* and TTNUs has made them the central hub of Russia's KMNS land rights recognition politics (Fondahl et al., 2001, p. 551).

In 2000, ten years after the first *obshchiny* were organized locally, federal legislators enacted the law "*On obshchiny*" (Russian Federation, 2000). The law recognized *obshchiny* as non-profit organizations (NPOs) and their economic activities solely for non-commercial purposes. The latter has been limited to a closed list of thirteen types of activities, including reindeer husbandry, hunting, and fishing (Russian Federation, 2000). The new legislation also required *obshchiny* created in the 1990s to change their status from commercial agricultural organizations to non-profit. Since the mid-2000s, the government has regularly stripped away the provisions of rights of *obshchiny* (Kryazhkov, 2015). The most critical of these, concerning the land rights of the *obshchiny*, were introduced by the new Land Code (Russian Federation, 2001a). The Code replaced the norm of land use "free of charge" with use "on lease." The new regulation eventually jeopardizes the very existence of the *obshchina*. No single *obshchina* can afford to pay even the minimal rent for thousands of hectares of land tenured under the restrictive conditions to use it only for non-commercial activities. Due to the municipal government reforms of 2004–2005, the self-governmental function of *obshchiny* at the local level also became invalid (Kryazhkov, 2015, p. 56).

The federal law FZ-49 defines territories of traditional nature use (TTNU) as "specially protected territories, established on the lands of *obshchiny* to ensure traditional nature use and preserve traditional ways of life" (Russian Federation, 2001). The legislator expels these territories from any property transfers (via buying-selling, lease, etc.). In the same vein as FPIC, the legislator recognizes the right of the KMNS to say no to industrial activities on such territory, yet without the veto power. If industrial activities in such an area are unavoidable, the law guarantees the affected communities compensation payments or land allocation elsewhere.

Since 1992, in many subjects (regions) of Russia, the authorities have established hundreds of TTNU under their jurisdiction (Tranin, 2010). Meanwhile, the federal government has failed to establish a single federal-level TTNU. Given the supremacy of the federal law, the future of regional-level TTNU remains peculiar. In the event of a potential conflict of national and regional jurisdictions, the latter would fail to protect the regional TTNU from being dismantled (Murashko and Rohr, 2018, p. 40).

As a cornerstone of FPIC, the participation and consultation of the KMNS affected by industrial activities are regulated by federal land, environment and subsoil legislation. The legislator requires companies to inform, consult and consider the local community's opinion regardless of their ethnic composition before implementing the project. The law provides two institutional channels of participation on the local level of governance of extractive developments for KMNS and non-KMNS people: an environmental review and public hearings.

The legislator obligates all developers to conduct *otsenka vozdeystviya na okruzhaiyschuiy sredu, OVOS* (comparable to an environmental impact assessment, EIA) (Russian Federation, 1995). This may include an *etnologicheskaya ekspertiza, EE* (comparable to a social impact assessment, SIA), but this is not obligatory. The results of an assessment of environmental impacts become subject to deliberation at a public hearing, a gathering where the community meets with developers and authorities to voice their concerns and expectations regarding the proposed activities. The public hearing ends with a protocol that includes these issues but has no legal force binding the company to implement them. While a public hearing implies a democratic and inclusive idea of governance, in practice, it gives the community only the tiniest degree of empowerment, making its participation through this channel rather a formality (Tulaeva et al., 2019).

To sum up, while the Russian legislation formally includes norms on participation, informing and consulting Indigenous peoples, the existing framework addresses FPIC neither entirely nor comprehensively. The Russian legislator's vision of the FPIC is narrow, as it impairs the fundamental importance of this principle to ensure Indigenous peoples' rights in the international legal framework (Kryazhkov and Garipov, 2019). Nevertheless, within the contemporary Russian federative state, numerous subjects (regions) provide better protection of KMNS rights than the corresponding federal law. One of the vanguard regions where regional lawmakers have made progress in incorporating the FPIC in KMNS legislation and its implementation is the Republic of Sakha (Yakutia) (Sleptsov and Petrova, 2019).

Contestation on FPIC in the RS (Ya): a case study

FPIC in the RS (Ya) legal framework

The Republic of Sakha, with an area of 3,084 million square kilometers, is one of three ethnic republics among the nine federal subjects of the Arctic Zone of the

Russian Federation (AZRF). The republic has a population of one million people, around half of whom have a Sakha (Yakut) ethnicity. The capital Yakutsk lies 4,900 kilometers east of Moscow. For centuries, for five ethnic groups, practicing a semi-nomadic way of life and closely connected to the land, these territories have been a homeland. According to the last census (2010), these groups include the Evens, the Evenki, the Dolgans, the Chukchi and the Yakagirs, making up just 4.2 percent of the region's population.

Sakha became part of the Russian Empire in the sixteenth century, and since then, its economic history has been one of resource exploitation (Tichotsky, 2000, p. 72). For Sakha's governors, ownership and control over the land (subsoils) have always been a matter of paramount importance. Within the Soviet command-administrative system, the republic's gold mining and diamond industries provided the national budget with significant foreign exchange earnings, ensuring its special status in relations with central authorities in Moscow (Tichotsky, 2000, p. 71). In early post-Soviet Russia, Sakha's elites successfully used land, indigeneity and ethnicity issues as resources in their negotiations with the federal center over land control, subsoil revenues and the strengthening of Sakha's sovereignty (Balzer and Vinokurova, 1996, p. 101).

Nicknamed "a storehouse of the country's diamonds, gold, tin, oil and gas reserves," RS (Ya) is also known for its protectionism toward KMNS through legislation and policy. The republic adopted most of the laws on KMNS earlier than the federal legislator (Table 5.1). These days the regional legislation provides better protection of KMNS rights than the corresponding federal legislation (Fondahl et al., 2020). Just a month after the presidential decree (1992) that recognized the land tenure rights of *obshchiny* and thus legitimized their inclusion in the debate on land privatization in the Russian North, Sakha politicians passed the regional *Obshchiny* Law (1992). During the next decades, the republic became the flagman in the organization of *obshchiny* and the territories of traditional nature use. These days it has 199 *obshchiny* and 62 TTNU, which comprise a significant share of such institutions in the Arctic Zone of the Russian Federation (Sakha Republic, 2020a).

Table 5.1 Legislation in the RS (Ya) and the Russian Federation on KMNS rights

Subject	RS (Ya) law adopted	Russian federal law adopted
Constitution	1992	1993
Obshchina KMNS	1992	2000
Reindeer Husbandry	1997	
Territories of Traditional Nature Use (TTNU)	2006	2001
Ethnological Expertise (EE)	2010	
Ombudsman For Indigenous Peoples' Rights (OIPR)	2013	
On Responsible Subsurface Resource Use	2018	

During the 2000s, political and administrative reforms sharply increased decentralization in the federal–regional relations, including the redistribution of tax flows, resource revenues and the unification of law. The federal legislator has failed to provide proper legal backing and guarantees to *obshchiny* and the TTNU. On the contrary, the expansion of the country's resource-based economy and the growth in energy demands around the world have led to numerous amendments and changes in federal legislation, further weakening the legal protection of the KMNS (Murashko and Rohr, 2015, p. 30).

In the Republic of Sakha, these processes have led to a "second wave" of regional lawmaking to strengthen control over the territory (subsoils) and promote good governance in KMNS affairs. Lawmakers' efforts have resulted in two enforcement mechanisms through ethnological expertise (EE) and the ombudsman for Indigenous peoples' rights (OIPR). Both instruments aim to compel companies to comply with international and national rules regarding information, consent and compensation for the KMNS affected by their industrial activities. In contemporary Russia, EE is an exclusive practice to RS (Ya), while the OIPR is limited to a few regions.

Even though norm obligating extractive companies to conduct EE was mentioned in the federal law two decades ago, legislators' efforts have not gone beyond the project stage (Novikova and Wilson, 2017). To fill this gap, in 2010, the RS (Ya) legislators issued a law on ethnological expertise. The law defines ethnological expertise as "a public service aimed to create conditions for meaningful dialogue and partnership between extractives and KMNS" and explicitly endorses the FPIC as its guiding principle (Sakha Republic, 2010).

Like a social impact assessment (SIA), ethnological expertise is a scientific study to measure planned industrial activities' cumulative impacts on the livelihood, culture and economies of the affected *obshchiny*. It results in a legal decision to support or reject the project, stating the amount of compensation that the company has to pay to the *obshchiny*. Unlike SIA's voluntary nature within what is comparable to an environmental impact assessment (OVOS), ethnological expertise is mandatory for all industrial activities planned in areas with *obshchiny* prior to implementing a project. Companies evading EE are subject to a fine. It is essential to emphasize that the binding character of ethnological expertise is limited only to territories of traditional nature use.

The Ombudsman for Indigenous Peoples' Rights (OIPR) is an independent government body that aims

> to institutionalize a guaranteed right of Indigenous peoples to have a special representative to advocate their interests in relations with authorities, businesses, and civil society organizations in a court and other settings.
>
> (Sakha Republic, 2013)

The OIPR is appointed by the Head of the RS (Ya) on the KMNS organizations' proposal. The mandate gives the ombudsman the authority to investigate KMNS complaints of maladministration and violation of their rights, exert non-judicial pressure to resolve conflicts involving the KMNS and submit annual recommendations to the RS (Ya) Parliament and its Head.

High stakes at Nezhda

Nezhda (*Nezhdaninskoye*) is the fourth-largest gold deposit in Russia (632 tons of ore reserves) located in the remote areas of the Verkhoaynsk mountain range in the northeast of RS (Ya) (Figure 5.1). The deposit was discovered in 1951, but due to global negative trends in gold prices and the economic crisis in 1998, the mine was closed. In 2015, Polymetal, one of the largest global gold producers, came to the RS (Ya) through the JSC South-Verkhoyansk Mining Company to restart Nezhda. Total capital expenditures for Nezhda are estimated at USD234 million, with a mine life of up to 2045 (Polymetal International plc, 2020).

Polymetal is an internationally active Russian precious metals public limited company registered in Jersey (UK). The company shows its commitment to corporate ethical conduct and responsibility through membership with the UN Global Compact, the Extractive Industries Transparency Initiatives (EITI) and the International Finance Corporation performance standards (IFC). Under its principal investor's requirements – the European Bank for Reconstruction and Development – the company undertakes to respect Indigenous peoples' land rights and integrate the FPIC in its operation.

Over a decade of operations in the Russian sub-Arctics, Polymetal has built the company's reputation responsive to Indigenous peoples' rights and environmental standards. Several national and international assessments have praised the company's

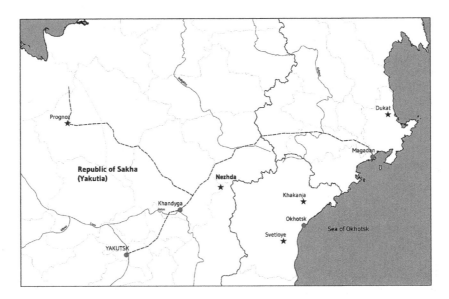

Figure 5.1 The Nezhda mine, the Republic of Sakha (Yakutia), Russia. ©Arctic Centre, University of Lapland.

environmental responsibility efforts and performance in Indigenous communities' engagements (Overland, 2016; Knizhnikov et al., 2018). A more detailed analysis of the company's social reporting shows that Polymetal does not have a specific corporate Indigenous policy and considers "Indigenous issues" among many other engagements with local communities. The company engages with local (Indigenous) communities through voluntary "in-kind" donations and philanthropy, rather than on a program basis. Moreover, the company does not have a formal grievance mechanism to provide affected Indigenous communities with access to remedy.

The area around Nezhda has high stakes not only for the company and mining. After the Tomponskyi state farm's liquidation, these plots were transferred to former workers, the Even reindeer herders, who organized their family-based *obshchina*. The *obshchina* received 396,000 hectares of land for forty-nine years as a usufruct (land tenure). On the cadastral passport that the *obshchina* has for the land, the plots are registered as hunting grounds, legitimizing their multi-purpose use for reindeer herding, hunting and fishing. However, the areas adjacent to Nezhda do not have the status of a territory of traditional nature use. Although the *obshchina* has applied to recognize these parcels as such, local authorities have rejected these applications, arguing that this can lead to a "conflict of interests" between different land (subsoil) users.

Since 2001, the *obshchina* has had a legal entity status as a non-profit organization of Indigenous peoples with reindeer husbandry as its principal activity. It owns a thousand reindeer, and its primary income comes from the republican subsidies for reindeer husbandry. Seventy percent of that small but stable income goes to herders' remuneration at USD 300 per month. The community is an active member of the republican branches of Indigenous peoples' and reindeer herders' associations, including the World Reindeer Herders Association (WRH), the Association of Indigenous Peoples of the North (AIPON) and the Union of the Nomadic Obshchiny (UNO).

The *obshchina* officially has eleven adult members registered as employees. Its organizational structure includes two brigades (camps), each led by brothers, while their sister, a well-known Even politician in the past, acts as chairwoman. The brothers and their families herd the deer and watch these remote territories all year round, whereas the chairwoman's job in Yakutsk is crucial to accessing the authorities, company headquarters and Indigenous associations to carry out necessary paperwork and networking. The combination of rural and urban members in the organizational structure and its strong ties with authorities and Indigenous associations ensure the *obshchina*'s access to various sites of negotiations, resources and flows (material and nonmaterial) regionally, nationally and internationally. Although these characteristics of the *obshchina*'s organizational capacity are not unique, they are also not typical of two hundred other *obshchiny* in the RS (Ya).

FPIC through the actors' contestation "talks" and "walks"

The data analysis revealed that the *obshchina* and Polymetal had different perceptions of FPIC. As a commercial entity, the company has viewed FPIC from a "minimalist" stance, narrowing its interpretation to national legislation and limiting its costs and responsibilities to affected *obshchiny* to only legally binding tasks.

In contrast, the community's perception of the FPIC is broad, based on the principles of reciprocity, mutual respect, shared responsibility and accountability. The actors' views influenced their contestation practices around two main areas; "consent" and "benefits-sharing."

The first area of disagreement between the *obshchina* and the company was about "consent," including who grants the consent in Nezhda and whose consent counts as legitimate. Polymetal entered the RS (Ya) in 2015, having signed cooperation agreements with the republican and Tomponskyi municipal authorities. The public hearing on Nezhda was held in the municipal center Khandyga, 250 km away from the mine. Most of the participants were representatives of the local authorities or the company and none of them informed the *obshchina* about the reopening of the mine or the hearings. The environmental impact assessment stated that the project would not affect any Indigenous *obshchina* and TTNU, and its environmental impact would be moderate. The hearing ended with a protocol supporting the Nezhda mine, which the company acknowledged as the local community's consent.

The *obshchina*'s normative stance concerning the "C" in the FPIC was different. Soon after the project started, the *obshchina* lost dozens of reindeer due to traffic accidents and shootings. These incidents and the "minimalist" conduct of Polymetal brought the *obshchina* chairwoman to the company's Yakutsk office to negotiate trade-offs. During the negotiations, the chairwoman challenged the legitimacy of the consent obtained, requiring the company to recognize the *obshchina* as one of the local consent-grantors. The chairwoman argued that the local consent, to be legitimate, must include the informed agreement of all those affected by the mining industry and, first of all, of "affected Indigenous communities." Voicing the "Indigenous" perspective in interpretations of FPIC as broad and inclusive, she used moral and non-legal character arguments, referring to customary law.

The company objected to this with its narrow interpretation of FPIC while using Russian legislation's normative language. The company claimed that the land around Nezhda was public property. The state granted the company a legal mining license. Even though the plots of the *obshchina* are adjacent to Nezhda, there is no legal recognition of these areas as TTNU. Consequently, the *obshchina*'s claims to the status of an "affected Indigenous community" lacked sufficient legal legitimacy. In turn, the *obshchina* insisted that even if their claims might have less legal significance without the official TTNU status, its demands to respect their rights and compensate for losses ultimately had moral legitimacy. How Polymetal respects these rights will have direct implications for its corporate reputation regionally, nationally and internationally.

The second area of contention between the *obshchina* and the company over the FPIC was benefits-sharing. Generally speaking, benefits-sharing implies distributing monetary and non-monetary benefits generated by implementing the development project and goes beyond compensations (Pham et al., 2013, p. 3). In Russia, benefits-sharing arrangements are not monolithic; their practice varies across legal regimes and institutional contexts of the regions (Tysiachniouk et al., 2018). In the RS (Ya), the engagement between Indigenous peoples and extractive companies regarding the distribution of benefits falls under two modes: quasi-formal bilateral agreement-making and formal agreement-making using ethnological expertise (EE).

As the above analysis shows, the legal framework limits the choices available to *obshchiny* if their territory does not have a TTNU status. The legislator excludes these *obshchiny* from the list of legitimate claimants for benefits-sharing through the EE. For comparison, the RS (Ya) hosts 418 extractive companies with 1,467 licenses to extract minerals, while only twenty-one ethnological expertise assessments were conducted in 2010–2020 (Sakha Republic, 2020b). The legislator left a large part of the *obshchiny* with a poor choice: to protest or negotiate with the company independently.

In the case under study, Polymetal argued its position on benefits-sharing from a commercial (minimalist) stance. The company justified its actions by Russian legislation, claiming its benefits-sharing with the RS (Ya) and the local municipality. These include taxes and revenues to the republican and local budgets, investments into infrastructure (building roads, electricity lines) and new jobs for the locals. According to the company, among other payments for 2016–2018, the company paid 27 million rubles only to the local budget. The municipality spent these to renovate a medical center, purchase computers for a school and celebrate Reindeer Herders' Day.

The *obshchina* objected, contesting the perceived legitimacy of these benefit-sharing arrangements as genuinely equitable. The chairwoman did acknowledge that the company's money had improved the residents' living standards in the municipal center. However, she emphasized that the reindeer herders in their remote camps received nothing from these "benefits" to somehow compensate for their damages, stress and risks. The chairwoman urged the company to provide a more targeted and justified distribution of benefits, ensuring the rights of affected reindeer herders to particular (and better) compensation.

Such interactions between the *obshchina*, Polymetal and the authorities, and their contestations around "local consent" and "benefits-sharing" are not unique to Sakha or Russia. The Russian "irregular governance triangle" (Petrov and Titkov, 2010) makes it a common practice on the ground for the authorities to go beyond the "intermediary" role and deliberately replace community (indigenous) voices, speaking on their behalf. Such a mode of interaction encourages companies to deal with the state's representatives instead of working with Indigenous *obshchiny* directly. The companies perceive "local consent" as an agreement with local authorities in exchange for social payments. The companies' money flows to capitals and municipal centers, while the Indigenous *obshchiny*, most affected by extractive activities, rarely enjoy these benefits. As already argued, the companies take a minimalist approach, limiting their costs and responsibilities to the affected *obshchiny* to tasks that are legally binding. The latter are few and easy to defy, given the principal role the extractive industries play in the country's resource-based economy and the deficit of the rule of law.

At the end of their first round of negotiations, the *obshchina* and Polymetal reached a verbal agreement that the company would pay damages for each deer killed. They also agreed to build a fence along the road to prevent deer–vehicle collisions. The deal was short-lived, and when the company failed to keep its promises, the *obshchina* submitted a complaint to the OIPR.

OIPR as a norm enforcer

The Indigenous peoples' right to use advocates to negotiate with more powerful counterparts in the FPIC process is recognized and broadly practiced. As international experience suggests, in contexts with a deficit of the rule of law and weakness of civil society organizations, Indigenous peoples have better chances to defend their rights with help from the specialized institution of the ombudsman (Krizsán, 2014). In Russia, the first institution of OIPR was established in the Krasnoyarsk region in 2008. Like other ombudsman-type institutions in Russia, the OIPR has a "personified" nature, the legitimacy and effectiveness of which heavily depend on political support from the regional authorities and civil society organizations (Bindman, 2017).

In the Republic of Sakha, the OIPR was established in 2014. During 2014–2019, the institution was led by Konstantin Robbek, who had extensive experience working with Indigenous rights in the republic as an activist, analyst and policymaker. A lawyer by education and Even by origin, he interned at the UN program for Indigenous practitioners on Indigenous advocacy and rights defense. For years of serving as the OIPR, Robbek has strengthened the new institution's capacity and mandate, not least with the support of local Indigenous organizations. The legitimacy and authority of the OIPR these days in the RS (Ya) is high and recognized by the extractives operating there.

In response to the *obshchina*'s complaint about Polymetal's misconduct, the OIPR organized a meeting between the parties to facilitate a dialogue. According to the ombudsman, the conflict situation between the *obshchina* and Polymetal was far from unique and had a standard set of characteristics and causes for such cases. At the core of the conflict was a lack of shared understanding of normative foundations of mutual conduct, rights and obligations between the parties in the context of extractive activities. Like every encounter between Indigenous peoples and extractives, the conflict manifested as an asymmetry of power, capacity and resources. Uncertainties, contradictions and numerous loopholes in federal legislation serve the companies' interests rather than protect Indigenous peoples' rights.

Given this background, the OIPR saw his role in balancing these power asymmetries by articulating challenges faced by the *obshchina* in legal terms and linking them to the powerful language of international law. Acting as a local normative-enforcer, the OIPR gave a broad interpretation of Indigenous peoples' rights, using relevant international standards (ILO 169 and UNDRIP) and referred to good examples of Indigenous–mining industry relations from other regions and countries.

Another crucial task of the ombudsman in mediating the conflict between the *obshchina* and Polymetal was to counteract the company's attempts to define and perform the FPIC solely on its own, following "minimalist" commercial visions. To do this, the OIPR leveraged its interpretative power and mandate as an institution affiliated with authorities to convince the company to accept broadly formulated interpretations of the FPIC process as authoritative.

While it is not always the case in practice, the mediation of the OIPR lifted the *obshchina*–Polymetal relations to a new level. One of the direct practical outcomes of the OIPR's facilitation was formalizing communication channels between the parties. The company appointed two officers to deal with the *obshchina*'s queries. Since then, communication has improved: it has become prompt, conducted by cell phone and respectfully. According to the Indigenous informants, a lack of respect and pervasive negative attitudes among the company's representatives had been some of the most significant barriers to building mutually trustful relations. Though these negative perceptions have not entirely disappeared, the facilitation of the OIPR has encouraged the company staff to progress with more sensitive and respectful attitudes toward the herders and their requests.

Soon after the meeting with the ombudsman, the *obshchina* and the company signed their first bilateral agreement. To date, the agreement practice is annual, bilateral, confidential and quasi-formal, offering benefits-sharing as "in-kind" services. For example, the company has subsidized a ten-kilometer-long fence along the main road. It regularly helps the herders to deliver food, fuel and equipment to their remote camps. Scrolling back on the history of their relationship with Polymetal, the members of the *obshchina* acknowledge the company's efforts to build positive mutual relations. Nevertheless, the current main concern of the *obshchina* remains to induce the company to step beyond its minimalist position toward more equitable benefits-sharing that will contribute to the *obshchina*'s long-term economic sustainability.

Conclusion

The case study of the *obshchina* in the Republic of Sakha (Yakutia) shows that the Russian Indigenous peoples' organizations, like their Arctic counterparts, increasingly recognize FPIC as a tool for empowerment. The analysis of the *obshchina*'s contestation practices highlights its agency (norm-generating power) to object and challenge the normative foundations of their relationships with the mining company and authorities, which they perceive as unjust, illegitimate and even immoral. As the study demonstrates, the availability and accessibility of institutional mechanisms to ensure *obshchiny* participation in deliberation forums is a matter of Indigenous peoples' success. The EE and the institution of the OIPR in the RS (Ya), complementing and enforcing each other, offer *obshchiny* different institutional doorways to broaden their participation in the governance of natural resources extraction at the local level. These mechanisms serve as the contestation sites, providing *obshchiny* with critical engagement with the norms to refine their rights' normative roots. Furthermore, the EE and the OIPR operate as local FPIC enforcers, which helps *obshchiny* enhance their rights to the FPIC and benefits-sharing. However, as the study shows, the interpretative power of the EE and OIPR is neither fixed nor conclusive and has its limitations.

The case study holds broader lessons for understanding the performance of FPIC on the ground that is not limited by the Russian extractive context. As extractive corporations' role in global governance grows, it is corporations rather than governments that take an increasingly leading role in promoting the FPIC. When the legislator does not require FPIC and does not control its implementation, it allows companies to independently decide what FPIC is about and where, how and to what extent it is to apply. As the case study shows, there is a risk that the company misuses the fundamental legal meaning of FPIC as the right of Indigenous peoples as it relates specifically to the *land* consent *prior* to any land disturbances (not afterward). Even when a company declares its commitment to FPIC, it often deprives the FPIC of its normative value, which is intended to enable self-determination of affected Indigenous *obshchiny* through true consultation and a share of the benefits to contribute to their sustainable development.

In the Russian context, FPIC can become a vehicle for Indigenous peoples to enable their right to self-determination in extractive developments but under specific provisions. These will require updating national legislation in line with international Indigenous peoples' rights supporting FPIC and empower the *obshchiny* through new, more democratic governance structures.

References

Balzer, M. and Vinokurova, U. (1996) "Nationalism, interethnic relations and federalism: the case of the Sakha Republic (Yakutia)," *Europe–Asia Studies*, 48(1), pp. 101–120.

Bindman, E. (2017) *Social rights in Russia: from imperfect past to uncertain future*. London: Routledge.

Deitelhoff, N. and Zimmermann, L. (2018) "Things we lost in the fire: how different types of contestation affect the robustness of international norms," *International Studies Review*, 22(1), pp. 51–76.

Fondahl, G., Lazebnik, O., Poelzer, G. and Robbek, V. (2001) "Native 'land claims,' the Russian style,". *The Canadian Geographer/Le Géographe canadien*, 45(4), pp. 545–561.

Fondahl, G., Filippova, V., Savvinova, A. and Shadrin, V. (2020) "Changing Indigenous territorial rights in the Russian North," in T. Koivurova, E.G. Broderstad, D. Cambou, D. Dorough and F. Stammler (eds.), *Routledge handbook on Indigenous peoples in the Arctic*. London: Routledge, pp. 127–142.

Gray, P. (2001) *The obshchina in Chukotka: land, property and local autonomy*. Halle: Max Planck Institute for Social Anthropology.

Jose, B. (2018) *Norm contestation: insights into non-conformity with armed conflict norms*. Cham: Springer International Publishing.

Hajer, M. and Wagenaar, H. (eds.) (2003) *Deliberative policy analysis: understanding governance in the network society*. Cambridge: Cambridge University Press.

Heinämäki, L. (2020) "Legal appraisal of Arctic Indigenous peoples' right to free, prior and informed consent," in T. Koivurova, E.G. Broderstad, D. Cambou, D. Dorough and F. Stammler (eds.) *Routledge handbook on Indigenous peoples in the Arctic*. London: Routledge, pp. 335–351.

Hoogensen Gjørv, G. and Goloviznina, M. (2012) "Introduction: can we broader our understanding of security in the Arctic?," in G. Hoogensen Gjørv, D. Bazely, M. Goloviznina, and A. Tanentzap (eds.) *Environmental and human security in the Arctic*. Abingdon, Oxon: Routledge, pp. 1–15.

Knizhnikov, A., Ametistova, L., Yudaeva, D., Markin, Y. and Dzhus, A. (2018) *Environmental transparency ration of mining and metals companies operating in Russia*. Moscow: World Wildlife Fund (WWF).

Kooiman, J., Maarten, B., Jentoft, S. and Pullin, R. (eds.) (2005) *Fish for life*. Amsterdam: Amsterdam University Press.

Krizsán, A. (2014) "Ombudsmen and similar institutions for protection against racial and ethnic discrimination," in T. Malloy and J. Marko (eds.) *Minority governance in and beyond Europe*. Leiden: Brill B.V. Nijhoff, pp. 61–84.

Kryazhkov, V. (2015) "Zakonodatel'stvo ob obshchinakh korennykh malochislennykh narodov Severa kak razvivayushchayasya Sistema," [Legislation on Indigenous obshchiny of the North as a developing system]. *Gosudatstvo i Pravo*, 1(11), pp. 49–59.

Kryazhkov, V. and Garipov, R. (2019) "Konventsiya MOT 169 i rossiyskoye zakonodatel'stvo o korennykh malochislennykh narodakh," [ILO Convention 169 and Russian legislation on Indigenous minorities] in N. Kharitonov and T. Gogoleva (eds.) *Analiz rossiyskoy i zarubezhnoy pravovoy bazy a takzhe pravopriminitel'noy praktiki v oblasti zashchity prav korennykh malochislennykh narodov Severa* [Analysis of the Russian and international legal framework and its implementation in the field of protection of the rights of Indigenous peoples of the North]. Moscow: Publication of State Duma, pp. 202–217.

Murashko, O. and Rohr, J. (2015) "The Russian Federation", in C. Mikkelsen and S. Stidsen (eds.) *The Indigenous world 2015*. Copenhagen: International Work Group for Indigenous Affairs (IWGIA), pp. 28–39.

Murashko, O. and Rohr, J. (2018) "The Russian Federation", in P. Jacquelin-Andersen (ed.) *The Indigenous world 2018*. Copenhagen: International Work Group for Indigenous Affairs (IWGIA), pp. 39–49.

Novikova, N. (2001) "The Russian way of self-determination of the Aboriginal peoples of the North: hypotheses for development," *Journal of Legal Pluralism*, 33(46), pp. 157–164.

Novikova, N. and Wilson, E. (2017) *Anthropological expert review: socio-cultural impact assessment for the Russian North*. Drag: ARRAN Lule Sami Centre.

Office of the High Commissioner for Human Rights, OHCHR (2018) *Role of the state and private sector in implementing the principle of free, prior and informed consent*. Available at: https://www.ohchr.org/Documents/Issues/IPeoples/EMRIP/FPIC/RussianFederation.pdf (Accessed April 21, 2020)

Ombudsman for Indigenous Peoples' Rights in the Republic of Sakha (Yakutia), OIPR (2020) Official website [Online]. Available at: https://iu-upkm.sakha.gov.ru (Accessed May 15, 2020)

Overland, I. (2016) *Ranking oil, gas and mining companies on Indigenous rights in the Arctic*. Drag: Arran Lule Sami Centre.

Peeters Golovizina, M. (2019) "Indigenous agency and normative change from 'below' in Russia: Izhma-Komi's perspective on governance and recognition," *Arctic Review on Law and Politics*, 10, pp. 142–164.

Petrov, N. and Titkov, A. (2010) *Irregular triangle: state–business–society relations*. Moscow: Carnegie Moscow Center.

Pham, T., Brockhaus, M., Wong, G., Le, N., Tjajadi, J., Loft, L., Luttrell, C. and Assembe Mvondo, S. (2013) *Approaches to benefit sharing: a preliminary comparative analysis of 13 REDD+ countries*. Bogor: Center for International Forestry Research (CIFOR).

Polymetal International plc (2020) Official website [Online]. Available at: https://www.polymetalinternational.com/ (Accessed May 15, 2020)

Rombouts, S. (2014) *Having a say, Indigenous peoples, international law and free, prior and informed consent*. Oisterwijk: Wolf Legal Publishers (WLP).

Russian Federation (1992) *Russian presidential edict of April 22, No 397. O neotlozhnykh merakh po zashchite mest prozhivaniya i khozyaystvennoy deyatel'nosti malochislennykh narodov Severa* [On urgent measures to defend the places of habitation and economic activity of the numerically small peoples of the North].

Russian Federation (1993) *Konstitutsiya Rossiyskoy Federatsii*. Adopted by national referendum on December 12, 1993 [The Constitution of the Russian Federation].

Russian Federation (1995) *Russian federal law of November 23, No 174-FZ. Ob ekologicheskoy ekspertize* [On ecological expertise].

Russian Federation (1999) *Russian federal law of April 30, No 82-FZ. O garantijah prav korennyh malochislennyh narodov Rossijskoj Federacii* [On guarantees of the rights of Indigenous numerically small peoples of the Russian Federation].

Russian Federation (2000) *Russian federal law of July 20, No 104-FZ. Ob obschikh printsipakh organizatsii obshchin korrennykh malochislennykh narodov Severa, Sibiri i Dal'nego Vostoka* [On general principles of organization of *obshchiny* of small-numbered peoples of the North, Siberia and the Far East of the Russian Federation].

Russian Federation (2001) *Russian federal law of May 7, No 49-FZ. O territoriyakh traditsionnogo prirodopol'zovaniya korennykh malochislennykh narodov Severa, Sibiri i Dal'nego Vostoka Rossiyskoy Federatsii* [On the territories of traditional nature use of the small-numbered peoples of the North, Siberia and the Far East of the Russian Federation].

Russian Federation (2001a) *Russian federal law of October 25, No 136-FZ. Zemel'nyy kodeks Rossiyskoy Federatsii* [Land Code of the Russian Federation].

Russian Federation (2010) *All-Russian Population Census 2010. Russian Federal State Statistics Service* [Online]. Available at: https://catalog.ihsn.org/catalog/4215/related-materials (Accessed September 9, 2020)

Russian Federation (2020) *Statistical data on non-governmental organizations in the Russian Federation* [Online]. Available at: http://unro.minjust.ru/NKOs.aspx (Accessed September 9, 2020)

Sakha Republic (2010) *Law of April 4, No 537-IV. Ob etnologicheskoy ekspertize v mestakh traditsionnogo prozhivaniya i traditsionnoy khozyaystvennoy deyatel'nosti korennykh malochislennykh narodov Severa Respubliki Sakha (Yakutiya)* [On ethnological expertise in places of traditional residence and traditional economic activities of Indigenous peoples of the North of Sakha (Yakutia)].

Sakha Republic (2013) *Law of June 24, No 1327-IV. Ob Upolnomochennom po pravam korennykh malochislennykh narodov Severa v Respublike Sakha (Yakutia)* [On Ombudsman for Indigenous Peoples' Rights in the Republic of Sakha (Yakutia)].

Sakha Republic (2020a) *Statistical data on obshchiny in the Sakha Republic* [Online]. Available at: https://arktika.sakha.gov.ru/TTP (Accessed September 10, 2020)

Sakha Republic (2020b) *Statistical data on ethnological expertise in the Sakha Republic* [Online]. Available at: https://arktika.sakha.gov.ru/_etnologicheskaja-ekspertiza_ (Accessed September 2, 2020)

Sirina, A. (2010). *Ot sovkhoza k rodovoy obshchine: sotsial'noekonomicheskiye transformatsii u narodov Severa v kontse XX veka* [From a state farm to a tribal community: socio-economic transformations among the peoples of the North at the end of the 20th century]. Moscow: IEN.

Sleptsov, A. and Petrova, A. (2019) "Ethnological expertise in Yakutia: the local experience of assessing the impact of industrial activities on the northern Indigenous peoples," *Resources*, 8, pp. 123–139.

Stammler, F. (2005) "The *obshchina* movement in Yamal: defending territories to build identities?," in E. Kasten (ed.) *Rebuilding identities: pathways to reform in post-Soviet Siberia*. Berlin: Reimer, pp. 109–134.

Stammler, F., Nysto, S. and Ivanova, A. (2017) *Taking ethical guidelines into the field for evaluation by Indigenous stakeholders*. Drag: ARRAN Lule Sami Centre.

Stammler, F. and Wilson, E. (2006) "Dialogue for development: an exploration of relations between oil and gas companies, communities and the state," *Sibirica*, 5(2), pp. 1–42.

Tichotsky, J. (2000) *Russia's diamond colony: The Republic of Sakha*. Amsterdam: Harwood Academic Publishers.

Tranin, A. (2010) *Territorii traditsionnogo prirodopol'zovaniya korennykh malochislennykh narodov Rossiyskogo Severa* [Territories of traditional nature use of Indigenous numerically small peoples of the Russian North]. Moscow: Institute of Government and Law, Russian Academy of Sciences.

Tulaeva, S., Tysiachniouk, M., Henry, L. and Horowitz, L. (2019) "Globalizing extraction and Indigenous rights in the Russian Arctic: the enduring role of the state in natural resource governance," *Resources*, 8(4), pp. 179–199.

Tysiachniouk, M., Henry, L., Lamers, M. and van Tatenhove, J. (2018) "Oil extraction and benefit sharing in an illiberal context: the Nenets and Komi-Izhemtsi Indigenous peoples in the Russian Arctic," *Society and Natural Resources*, 31(5), pp. 556–579.

Wiener, A. (2014) *A theory of contestation*. Berlin: Springer.

Wiener, A. (2017) "Agency of the governed in global international relations: access to norm validation," *Third World Thematics: A TWQ Journal*, 7(5), pp. 1–17.

6 The role of the Tłįchǫ Comprehensive Agreement in shaping the relationship between the Tłįchǫ and the mining industry in the Mackenzie Valley, Northwest Territories (NWT), Canada

Horatio Sam-Aggrey

Introduction

The 1973 Calder decision by the Supreme Court of Canada on Aboriginal title to land thrust the unsolved issues of land claims, Aboriginal title and self-government into Canada's political consciousness (See Calder v. Attorney General of B.C. 1973). The decision, which established that Aboriginal title pre-existed the assertion of British sovereignty in British Columbia (McKee 2009), marked the beginning of a new era of government–Aboriginal relations in Canada. The recognition and affirmation of Aboriginal rights, in Section 35 of the Constitution Act, 1982, further bolstered the status of land claims.

In response to the Calder ruling, the federal government instituted a land claims policy that would encourage the negotiation and settlement of modern treaties between the Crown and Canada's First Nations. Modern treaties or comprehensive land claims generally arise in areas of Canada where Aboriginal land rights have not been dealt with by treaties or through other legal means. These treaties are negotiated between the Aboriginal group, Canada and the province or territory. Since 1973, Canada and its negotiation partners have signed 26 comprehensive land claims agreements (Government of Canada 2015), covering over forty percent of Canada's land mass (Land Claims Agreements Coalition n.d.).

While each land claim is different, these agreements generally cover ownership, use and management of land, and they clarify how renewable and non-renewable resources will be owned, managed and regulated. They also outline measures to promote economic development and protect Aboriginal culture (Government of Canada 2015). Drawing on available literature on the relationship between the Tłįchǫ and the mining industry, a critical analysis of documents, including the Tłįchǫ Agreement, co-management board documents, academic literature and twelve semi-structured interviews with officials of the Tłįchǫ Government and co-management boards, this chapter examines the role of the agreement in shaping the relationship between the Tłįchǫ and the mining industry in the Mackenzie Valley of the NWT of Canada.

The main research question of this chapter is: How does the Tłįchǫ Agreement structure the interaction between the Tłįchǫ Government and the mining industry in the resource governance framework in the Mackenzie Valley? A secondary

DOI: 10.4324/9781003131274-6

research question is: Does the Tłı̨chǫ Agreement provide political leverage for the Tłı̨chǫ Government in its relationship with the mining industry?

The primary focus of the chapter is on Tłı̨chǫ participation in the resource management regime in the Mackenzie Valley and the role of impact benefit agreements (IBAs) in structuring the relationship between the Tłı̨chǫ and the mining companies. In this light, two key provisions of the agreement come into focus—the provisions enabling the Tłı̨chǫ to participate in the resource co-management process in the area (Chapter 22) and Chapter 23, which requires proponents to negotiate IBAs covering major mining projects with the Tłı̨chǫ.

I will first summarize the main provisions of the Tłı̨chǫ Agreement related to the mining industry and will then outline specific ways that the Tłı̨chǫ Government interacts with the mining industry as a directly impacted party in the resource co-management governance structure in the Mackenzie Valley. Third, this chapter assesses the role that IBAs play in shaping the interaction between the Tłı̨chǫ and the mining industry in the area. Finally, the chapter outlines the innovative governance structures emanating out of the provisions of the IBAs and the Tłı̨chǫ Agreement with concluding statements.

Tłı̨chǫ land claims and self-government agreement

The Tłı̨chǫ people, also known as the Dogrib, are an Aboriginal people that have used and occupied lands in and adjacent to the NWT of Canada since time immemorial. The Tłı̨chǫ Land Claims and Self-government Agreement, signed by the Dogrib Treaty 11 Council, the Government of Canada and the Government of Northwest Territories (GNWT) on August 25, 2003, came into effect on August 4, 2005 (Government of Canada 2005). This section highlights some of the important provisions of the Tłı̨chǫ Agreement which have significantly impacted the Tłı̨chǫ's relationship with the mining industry. The provisions outlined here are by no means the only ones in the agreement that are impacting the ways in which the Tłı̨chǫ interact with mining companies, but they represent the most significant provisions structuring the ways in which those interactions take place.

An important provision of the agreement is the clarity that it provides about Tłı̨chǫ ownership of 39,000 square kilometers of land between the Great Slave Lake and the Great Bear Lake in the NWT (see Figure 6.1), including subsurface and surface rights, and jurisdiction over land and resources (Government of Canada 2005).

The agreement also includes provisions for Tłı̨chǫ participation in the resource management process in the Mackenzie Valley through their participation in co-management boards, and a share of mineral royalties from developments in the area (Government of Canada 2005).

Another major provision of the agreement is the creation of the Tłı̨chǫ Government, with lawmaking authority over Tłı̨chǫ citizens in their communities and on their lands. This lawmaking authority includes aspects of education, adoption, child and family services, training, income support, social housing and Tłı̨chǫ language and culture (Government of Canada 2005). Importantly, with the formation of the Tłı̨chǫ Government, the Tłı̨chǫ have become an influential decision-maker and resource manager in its traditional area.

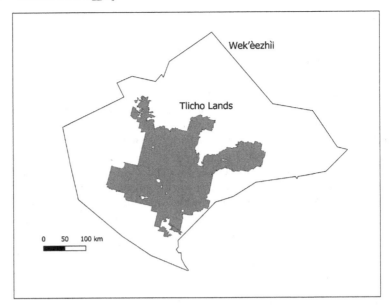

Figure 6.1 Map showing the areas covered by the Tłįchǫ Agreement. ©Arctic Centre, University of Lapland.

The Tłįchǫ Agreement also requires proponents of major mining projects to consult with the Tłįchǫ Government in order to develop IBAs covering their activities (Government of Canada 2005). An impact benefit agreement is a general term to describe a negotiated agreement between project proponents and Aboriginal communities to mitigate the various social, economic and biophysical impacts on one or more Aboriginal groups (Hitch 2006; Fidler and Hitch 2007). These negotiated and mainly private contractual agreements outline mitigation measures and the benefits that a local community can expect from the development of a local resource in exchange for its support and cooperation.

These provisions of the Tłįchǫ Agreement have provided the foundations for systematic and increased Tłįchǫ participation in the resource management process in the Mackenzie Valley. The next section outlines some of the processes put in place by the agreement which have facilitated the participation of the Tłįchǫ in resource management in the area.

Tłįchǫ Agreement and resource governance in the Mackenzie Valley

In 2019, the mining industry accounted for approximately 27 percent of the NWT's gross domestic product (GDP) (GNWT 2020a), making the extractive industry an important pillar of the economy of the territory. Environmental assessment (EA), which is the process of identifying, predicting, evaluating and mitigating the environmental, social and other relevant effects of development proposals prior to major decisions and commitments, is an important component of government oversight of new and operating mines. The EA process in the NWT is based on a co-management approach, rooted in the legal and cultural frameworks of comprehensive land claims agreements (CLCAs).

There are three settled CLCAs in the Mackenzie Valley: the Gwich'in (1992), the Sahtu (1994) and the Tłıcho (2005) (see Figure 6.2). The Government of Canada enacted the Mackenzie Valley Resource Management Act (MVRMA) in 1998 to implement the federal government's land claim obligations to the Gwich'in and Sahtu peoples (Government of Canada 1998). This legislation implements the EA sections of the agreements. The law was amended in 2005 to accommodate the Tłıcho Agreement.

The enactment of the MVRMA required the establishment of independent co-management boards to run the various stages of the EA and the regulation of new and existing extractive activities. These boards, which are responsible for various management processes from wildlife to water to EA, are referred to as "co-management" boards because they are made up of equal representatives of Indigenous communities and government representatives (territorial and federal governments). They function as decision-making bodies that are responsible for

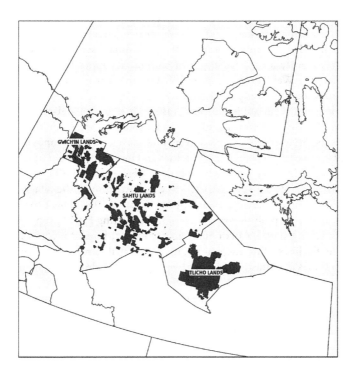

Figure 6.2 Map showing the areas covered by the three comprehensive land claims agreements in the Mackenzie Valley.

Source: Government of Canada n.d. (https://open.canada.ca/data/en/dataset/be54680b-ea62-46f3-aaa9-7644ed970aef#wb-auto-6). ©Arctic Centre, University of Lapland.

the day-to-day management of resources in their settlement areas, though in some cases the Minister of the Environment (federal or territorial) retains ultimate decision-making authority.

In 2005, as per the MVRMA and the Tłįchǫ Agreement, the Wek'èezhìi Land and Water Board and the Wek'èezhìi Renewable Resources Board were established as co-management decision-making boards to manage the resources in the Wek'èezhìi region. The membership of these boards is shared equally by the Tłįchǫ Government (fifty percent of members) and by the Government of Canada and the Government of the Northwest Territories (fifty percent of members).

The Wek'èezhìi Land and Water Board is responsible for preliminary screenings of developments and can refer projects to EA or environmental impact review in their jurisdictions. Preliminary screening is the first stage of environmental impact assessment required for most proposed developments. This stage examines a proposed development to decide if it might cause significant adverse impacts on the environment or if it might cause significant public concern. If so, it will be referred to environmental assessment. If not, the development proceeds directly to the regulatory phase.

The Wek'èezhìi Land and Water Board also regulates the use of land and water and the deposit of waste through its authority to issue, amend, suspend and renew land use permits and water licenses throughout the Tłįchǫ traditional territory. Of all projects proposed in the Mackenzie Valley, around 95 percent only go through the preliminary screening stage, and only about five percent are referred to the EA stage (Ehrlich 2016). For minor projects, the preliminary screening stage represents the only opportunity for the inclusion of community input before the regulatory phase. Hence, the Wek'èezhìi Land and Water Board plays an important part in regulating new and operating industries.

The Wek'èezhìi Renewable Resources Board is responsible for managing wildlife and wildlife habitat (forests, plants and protected areas) in the Wek'èezhìi region. The board also pre-screens and reviews development proposals and applications and collaborates on research activities and programs related to wildlife. This board is a mainly advisory body, whose powers derive from the Tłįchǫ Agreement.

The enactment of the MVRMA led to the formation of the Mackenzie Valley Environmental Impact Review Board (MVEIRB) in 1998. This board is responsible for reviewing all preliminary screenings, conducting EA of proposed developments and creating panels to conduct environmental impact reviews if necessary. Based on the findings of its assessments, the board makes recommendations to the responsible GNWT minister on whether a proposed development proceeds to regulatory review or not.

Having highlighted the institutions created to implement sections of the Tłįchǫ Agreement, it is vital to outline in more granular details some of the mechanisms through which the Tłįchǫ Government and community participate in the processes and how that shapes their relationship with the mining industry.

Tłįchǫ Government's role in resource management in the Mackenzie Valley

Primarily, the Tłįchǫ's participation in the resource management process and their interaction with the mining industry in the Mackenzie Valley is largely dictated by the provisions of the MVRMA, the Tłı̨chǫ Agreement and Section 35 of the Canadian constitution. This participation framework is operationalized through the processes led by the co-management boards.

In addition, the IBAs signed between the Tłı̨chǫ Government and respective mining companies provide avenues for Tłı̨chǫ input in resource governance in the Mackenzie Valley. Impact benefit agreements are sanctioned by the Tłı̨chǫ Agreement, but they are largely delinked from the co-management board processes because they are confidential private agreements between the companies and the Tłı̨chǫ Government.

In the comment below, an employee of the Tłı̨chǫ Government and a Tłı̨chǫ elder sums up the impact of the Tłı̨chǫ Agreement on Tłı̨chǫ participation in resource management in their traditional territory.

> Look, you (industry) had the freedom to do whatever you wanted up until when we had the agreement. Then it was like a God-given right for mining companies to come and do whatever they like. They have a legacy of leaving uncleaned and unattended sites. That is the freedom that they had and that is the kind of freedom that they are used to. Now that we have our agreement (Tłı̨chǫ Agreement) our views and knowledge are going to have to be taken into consideration because we are part of the system of approvals.

Tłı̨chǫ participation in resource management takes many forms, including membership of resource co-management boards, participation in the mine regulatory process as a directly impacted stakeholder and negotiating IBAs with the mining companies (IBAs will be discussed later in this chapter).

Tłı̨chǫ membership of co-management boards

The resource co-management boards, which are responsible for regulating development projects, are institutions of public government, with rights and responsibilities that are at the forefront of independent administrative decision-making. They are granted the powers, rights and privileges of a superior court with respect to examination of witnesses and the production and inspection of documents (Government of Canada 1998). Board decisions are formally made by majority vote but in practice are almost always by consensus (Government of Canada 1998).

One way that the Tłı̨chǫ Government has influenced the resource management process is through representation on co-management boards (they appoint fifty percent of appointees to the Wek'èezhìi Land and Water Board as well as Wek'èezhìi Renewable Resources Board) and they have a representative on the Mackenzie Valley Environmental Impact Review Board (where Indigenous members constitute half of the members of the board).

Tłı̨chǫ representatives on the boards are usually from the communities impacted by mining and are often hunters, trappers or fishers, who know their territory extensively and often possess traditional ecological knowledge about the area. Results from the interviews indicate that Aboriginal representation on the boards plays an important role in facilitating the incorporation of Indigenous local knowledge in the co-management process. An employee of one of the boards commented on the importance of having Aboriginal representatives on the board:

> One reason why board members are appointed is because they have extensive knowledge of the area that they are from. During an exchange over caribou migration routes around Ekati mine between a representative of the developer and a community member at one of our public hearings, a board member stood up and provided information that backed the traditional knowledge provided by the community. He walked up to the screen and commented "we used to hunt at that particular location but the caribou don't go there any-more. The caribou go up here, not down here" (pointing to a location on the map). We have situations where board members are making decisions on resource development in areas where they trapped or hunted, so they have a lot of knowledge about those areas.

Through its appointees in the co-management boards, especially the Wek'èezhìi Land and Water Board and the Wek'èezhìi Renewable Resources Board, the Tłįchǫ Government is able to exert some influence in the impact assessment of mining projects in the Mackenzie Valley.

Tłįchǫ participation in EA and the regulation of mines as a directly impacted stakeholder

There are three possible levels of EA before a project is approved and goes through the regulatory process (Figure 6.3). Section 114 of the Mackenzie Valley Resource Management Act states that the assessment entails preliminary screening, EA and an environmental impact review. Depending on the complexity of the project, a

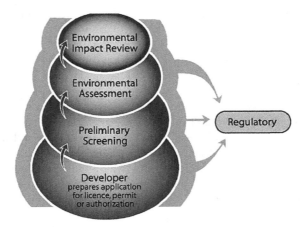

Figure 6.3 Diagram showing the different stages of EA process in the Mackenzie Valley.

Source MVERIB (used with permission of the MVEIRB).

development proposal may have to go through one or two or all of the stages before proceeding to the regulatory process.

Apart from its representation on the co-management boards, the Tłıchǫ Government is also an active participant in the impact assessment of projects in their territory. The MVEIRB and the Wek'èezhìi Land and Water Board consider the Tłıchǫ as a directly impacted stakeholder in preliminary screening, EA and other regulatory processes, a designation that gives the Tłıchǫ the opportunity to scrutinize projects going through the EA and regulatory processes.

Tłıchǫ interaction with the mining industry in preliminary screening

During preliminary screening, the Wek'èezhìi Land and Water Board's evaluates whether projects "might have significant adverse impacts on the environment, or might cause public concern" (Mackenzie Valley Environmental Impact Review Board, MVEIRB 2017). If so, the board refers the projects to EA for more scrutiny. Otherwise, the application may continue through permitting and licensing stages (Mackenzie Valley Environmental Impact Review Board, MVEIRB 2017).

When a project is in preliminary screening in the Wek'èezhìi region, the Tłıchǫ Government is asked, as the representative of directly impacted people, to identify potential environmental and socioeconomic impacts of the proposed development and make recommendations for minimizing or mitigating these impacts. The Wek'èezhìi Land and Water Board then reviews the Tłıchǫ Government's contributions and in many cases incorporates some or all of their concerns and suggestions for mitigation in their final decision on the preliminary screening.

Significantly, as the representative of the Tłıchǫ, the MVRMA also empowers the Tłıchǫ Government to refer a development project to the review board for an EA, even if the preliminary screening decision by the Wek'èezhìi Land and Water Board did not do so (see s.126(2), (3) & (4) of the MVRMA). This suggests that the concerns of the Tłıchǫ Government are given considerable weight in the regulatory process.

Tłıchǫ interaction with the mining industry during environmental assessment

When projects that can potentially negatively impact the Tłıchǫ are referred to the EA stage, the Tłıchǫ Government typically participates in most of the stages in the process. This includes issue scoping, review of the draft and final terms of reference; review of the environmental impact statement and supplemental information; generating, and possibly responding to, information requests and participation in technical meetings and public hearings.

In all these stages the Tłıchǫ Government provides its input on diverse issues related to the projects' impacts on the lives of its people. In some instances, they request more information on the project from the proponent or make recommendations to the MVEIRB and proponents on ways of mitigating negative impacts of the projects. The proponents often respond to the Tłıchǫ Government's information requests and often modify project plans based on the government's recommendations.

During the consultation process for Dominion Diamond's Jay Project, several representatives of Aboriginal governments and organizations (including the Tłıchǫ Government) recommended that the MVEIRB require the proponent to convene and fund an Aboriginal traditional knowledge elders group to assist in monitoring caribou behavior and recommend ways of mitigating impacts on caribou migration during the construction, operations and closure of the mines. This recommendation was accepted by the review board and was included as a measure (a requirement that must be fulfilled before the project commences) (Measure 6–5) in the board's final decision paper (Mackenzie Valley Review Board, MVRB 2016). Importantly, this requirement was included in the IBAs signed between Dominion Diamond and the Aboriginal governments and organizations.

Tłıchǫ Government's role in regulatory approvals

When an application for a land use permit or water license is submitted to the Wek'èezhìi Land and Water Board by a mining company in the post-EA phase, the Tłıchǫ Government is also given the opportunity to review and identify potential impacts and recommend mitigations to identified impacts. The government assesses potential impacts of the project and provides its comments to the Wek'èezhìi Land and Water Board, who in turn asks the proponents to address the comments and provide a response on how they intend to address the Tłıchǫ Government's concerns.

In issuing the licenses, the Wek'èezhìi Land and Water Board may incorporate the recommendations of the Tłıchǫ Government as "conditions" in the licenses of the developers. Conditions are legally binding provisions, which the proponents are required to follow in order to maintain their license. For example, in response to Dominion Diamond's Ekati Jay mining project, a technical intervention by the Tłıchǫ Government requested that the Wek'èezhìi Land and Water Board impose a condition on the land use permit requiring the company to seek the advice of Aboriginal elders on the location, design and operation of caribou crossings on the Jay Road, esker crossing and waste rock storage area egress ramps to limit the impacts to caribou mobility (WLWB 2017). This condition was accepted by the board, and Dominion Diamond is required by law to abide by this condition to maintain their land use permit.

It is evident that the Tłıchǫ Agreement has had a positive impact on the role of the Tłıchǫ people in their settlement area's economy and their relations with industry, as well as ensuring that they have a meaningful and effective voice in land and resource management decision-making. Furthermore, the agreement has helped to protect the Tłıchǫ's traditional way of life. Impact benefit agreements are another mechanism through which the Tłıchǫ participate in managing the impacts of mining projects on their traditional territories.

Comprehensive land claims agreements, corporate social responsibility and impact benefit agreements

Recent Supreme Court of Canada decisions require that the federal and/or provincial Crown consult with and, where appropriate, accommodate Aboriginal groups when intending to engage in activities that could adversely affect their potential or established Aboriginal and treaty rights (Haida Nation v. British Columbia (see Haida Nation v. Minister of Forests of British Columbia 2004), Taku River Tlingit First Nation v. British Columbia (see Taku River Tlingit First Nation v. British Columbia 2004) and Mikisew Cree First Nation v. Canada (see Mikisew Cree First Nation v. Canada 2005). Although the legal duty to consult rests with the Crown, procedural elements of consultation are often delegated to industry proponents for specific projects (Government of Canada 2011).

While in most cases (especially outside of the northern territories) proponents have no legal obligations to consult and accommodate, they often use IBAs to secure local support and reduce the likelihood of legal action on the basis of inadequate consultation.

An impact benefit agreement (IBA) is a general term to describe a negotiated agreement between project proponents and Aboriginal communities to mitigate the various social, economic and biophysical impacts on one or more Aboriginal groups (Hitch 2006; Fidler and Hitch 2007). These agreements serve to address the potentially adverse effects of development activities on Aboriginal communities, while also providing benefits or some compensation for these activities. They also provide the project proponent with the social license to operate from the affected Aboriginal community's perspective. Other common terms for IBAs are benefit sharing agreements, project support agreements, participation agreements, accommodation agreements and consultation agreements.

The contents of IBAs tend to be context-specific, depending on a wide variety of factors. These factors include the rights and interests at issue, the unique provisions of each land claims agreement, whether the Aboriginal beneficiaries' land tenure rights are surface, subsurface or both, the extent of perceived effects from the proposed project on the land, the project parameters and the bargaining power and leverage of the Aboriginal groups (Hitch 2006; Gibson et al. 2011).

As a result of the land claims agreements in the region, industry proponents in the Mackenzie Valley are obliged to negotiate with Aboriginal government or organizations to conclude an IBA in advance of mineral development. According to section 23.4.1 of the Tłı̨chǫ Agreement, proponents of major mining projects that require Government authorization and that will have an impact on Tłı̨chǫ citizens are required to enter into negotiations with the Tłı̨chǫ Government for the purpose of concluding an agreement relating to the project. They are required to consult with the Tłı̨chǫ Government on the potential social, economic, cultural and environmental impacts of the project on Tłı̨chǫ.

In line with this provision, the Tłı̨chǫ Government has signed four sets of IBAs with the diamond mines operating in the NWT (see Table 6.1), that is, the Ekati, Diavik, Gahcho Kué and Snap Lake mines. Each of these agreements is confidential and establishes a mechanism for mitigating the negative impacts of the mines' activities, while enhancing the benefits of the mine.

The details of the Tłı̨chǫ IBAs are confidential but generally involve the provision of funding for skills training, employment and business opportunities for Tłı̨chǫ businesses. Mine training and education, environmental and cultural protection provisions (land programing) also feature prominently. Each agreement calls for the

Table 6.1 List of signatories to the IBAs in the Mackenzie Valley

Project	Proponent	Indigenous organizations (and dates IBAs were signed)
Ekati Diamond Mine	BHP	Tłı̨chǫ Government (then Dogrib Treaty 11) (Oct. 1996); Lutselk'e Dene First Nation (Nov. 1996); Yellowknives Dene First Nation (Nov. 1996); North Slave Métis Alliance (Jul. 1998); Kitikmeot Inuit Association (Dec. 1998)
Diavik Diamond Mine (began production in Jan. 2003)	Diavik Diamond Mines Inc. (Aber Diamonds and Rio Tinto plc)	North Slave Métis Alliance (Mar. 2000); Tłı̨chǫ Government (then Dogrib Treaty 11) (Apr. 2000); Yellowknives Dene First Nation (Oct. 2000); Kitikmeot Inuit Association (Sept. 2001); Lutselk'e Dene First Nation (Sept. 2001)
Snap Lake Diamond Mine (began production in Jul. 2008)	De Beers Canada	Yellowknives Dene First Nation (Nov. 2005); Tłı̨chǫ Government (Mar. 2006); North Slave Métis Alliance (Aug. 2006); Lutselk'e and Kache Dene First Nation (Apr. 2007)
Gahcho Kué (commenced commercial production in March 2017)	De Beers Canada Inc. (De Beers) and Mountain Province Diamonds	North Slave Métis Alliance (Jul. 2013); Tłı̨chǫ Government (January 2014); Yellowknives Dene First Nation (Feb. 2014); Lutsel K'e and Kache Dene First Nation (Jul. 2014) NWT Métis Nation (Dec. 2014); Deninu Kué First Nation (Dec. 2014)

Source: Levitan 2012 and Debeers Group n.d.

creation of a joint implementation committee to outline responsibilities, tasks and timelines to reach project-related employment and business development targets.

These agreements, which are negotiated outside of the EA process, cover issues which are normally identified in the EA process and during the interactions between the proponent and the Tłı̨chǫ Government. Although IBAs are separate from the regulatory processes, the lines between the two can sometimes be blurred, with each process impacting the other.

The IBAs and the structures emanating from them serve an important liaison function between the Tłı̨chǫ and the mining companies, by providing opportunities for Tłı̨chǫ participation in identifying and mitigating the impacts of mining activities.

For example, the impact of mining on the caribou population is of particular concern to the Tłįchǫ. A Tłįchǫ Government employee notes the importance of caribou to Tłįchǫ culture and the role of the government in protecting the caribou:

> Caribou is everything. Caribou provides you with food, clothes and shelter. People had to be good hunters to survive. They had to be good at tanning the hides so that they have shelter and clothes. When it comes to marriage, women would want to marry a very good hunter, while men would want to marry someone who knows how to work the caribou skin so that you survive. So now that we have our own government and agreement (reference to comprehensive land claims agreement), it is the responsibility of this government to make sure that the caribou is not destroyed and that is what they are doing with all of these Tłįchǫ knowledge projects.

Given that impact benefit agreements are mandated by the Tłįchǫ Agreement, the Tłįchǫ's political leverage while negotiating the IBAs emanates from legal provisions of the land claims agreement. Two interviewees from the Tłįchǫ government conceded that the Tłįchǫ Agreement had strengthened the Tłįchǫ Government's bargaining power during negotiations of the IBA.

The Tłįchǫ IBAs include provisions to support ongoing research, monitoring and mitigation mechanisms for the impacts of mining on caribou. The Tłįchǫ Government, with IBA funding, conducts various traditional ecological knowledge and land use studies to inform their technical submissions to the co-management boards. Knowledge attained from this program is also submitted to the mines for incorporation into their management plans in the construction, operation and reclamation phases.

In addition, in some cases, regulatory requirements (especially those related to traditional ecological knowledge) imposed on the proponent of a project by the co-management boards are funded through the IBAs. For example, in its final EA report for Dominion Diamond's Jay Project, the MVEIRB required the proponent to convene and fund an Aboriginal traditional knowledge elders group to assist in monitoring caribou behavior and recommend ways of mitigating impacts on caribou migration during the construction, operations and closure of the mines. This group was formed and its activities are being funded through the IBAs.

Since impact benefit agreements are confidential and hard to get information on, socioeconomic agreements are used as proxies for IBA content. These are agreements between the territorial government, Aboriginal governments and organizations representing potentially impacted communities, and the proponents of major projects (GNWT 2018). Socioeconomic agreements typically include company commitments to employment and business opportunities, cultural and community wellbeing. The GNWT oversees the implementation of these agreements, tracking progress on the goals of the socioeconomic agreements and coordinating its efforts under each agreement. There are five such agreements in the NWT (GNWT 2018). Socioeconomic agreements are essentially an amalgamation of all the IBAs signed with all the Indigenous governments and organizations in the Northwest Territories.

Hence their contents largely mirror the contents of IBAs except for specific contextual provisions and direct monetary payments made to Indigenous communities.

In fulfilling the requirements of the IBAs and the Tłı̨chǫ Agreement, the Tłı̨chǫ Government engages with the diamond mines industry in different ways and on various issues related to the operation and impacts of the mine. The mines hold numerous consultation meetings with potentially affected Indigenous groups (including the Tłı̨chǫ Government) to share project information, discuss findings of environmental baseline studies and solicit traditional knowledge related to the mine. They also facilitate site visits to the mine's site for potentially affected Indigenous groups. Regularly scheduled meetings between senior management and Tłı̨chǫ Government representatives, and community meetings on issues of importance to the communities have been key to building a strong working relationship between the Tłı̨chǫ and the mines.

Employment and training are central provisions of many IBAs including the Tłı̨chǫ Government's agreement with the diamond mines. The Tłı̨chǫ government provides the names of potential employees for mines, trainees for mine-related skills and training, and the names of Tłı̨chǫ -owned companies for provision of various goods and services to the mines.

Another example of a direct engagement between the two parties is the joint Tłı̨chǫ Government and Dominion Diamond mine project to examine if dust from the mining activities is absorbed into the lichen and ingested by caribou. The project sampled soil and lichen at specific distances from the mine site and then investigated how dust from the mine potentially affected caribou use of the area. The study used traditional knowledge (record of past caribou use of the area) and modern scientific methods to sample the soil and lichens (Tłı̨chǫ Research and Training Institute 2018).

While much of the engagement with the mining proponents is direct one-on-one collaboration on diverse issues, the Tłı̨chǫ Government also engages with industry in multi-stakeholder organizations (usually consisting of regulators, industry and other Indigenous groups who have signed IBAs related to the operations of the mines). One such organization is the Environmental Monitoring Advisory Board (EMAB) set up to oversee the operations of the Diavik Mine.

The Diavik Diamond mine, situated some 200 kilometers south of the Arctic Circle in the NWT, at the bottom of Lac de Gras, commenced production in 2003. One of the conditions for the approval of the mine by the federal government was the signing of an environmental agreement, intended to ensure the environment around the Diavik Mine site is protected. It came into effect in March 2000, several months before the mine was given government approval to begin construction (Environmental Monitoring Advisory Board, EMAB n.d.). The Environmental Monitoring Advisory Board was constituted under the environmental agreement for the Diavik Diamond Project.

The parties to the environmental agreement are all the First Nations communities who signed IBAs with Diavik Mine (Tlicho Government, Lutsel K'e Dene First Nation, Kitikmeot Inuit Association, North Slave Métis Alliance and the Yellowknives Dene First Nation), Diavik Mine, Government of the Northwest Territories, Government of Canada and the Government of Nunavut. Each party to the agreement appoints a regular member and an alternate to sit on the EMAB, which has majority representation by the five Aboriginal groups who have signed an IBA with the Diavik. The other Aboriginal communities are stakeholders because their territories also border the mine.

The EMAB provides oversight of Diavik Diamond mine and the regulatory process to ensure protection of the land, water, air and wildlife in the Lac de Gras area. The board also monitors the regulators and managers and makes recommendations to them, and may also make recommendations when regulators raise issues, or when regulators and Diavik disagree on an issue. These recommendations may also be forthcoming when either Diavik or the regulators do not address a question that the board thinks is important. The regulators and Diavik Mine must accept the board's recommendations or give reasons why they did not (Environmental Monitoring Advisory Board, EMAB n.d.).

The EMAB also undertakes community-based monitoring to help communities better understand the interaction between the mine and the environment. The board's traditional knowledge panel is mandated to work with local communities and assist the board in ensuring that Aboriginal knowledge is appropriately and meaningfully incorporated into the planning and management of the mine. In addition to providing comments on reports submitted by the diamond mine and the regulators, the Tłıchǫ Government provides capabilities on traditional knowledge about caribou to the board (EMAB 2019).

As an independent public watchdog of Diavik, the EMAB works at arm's length from Diavik and the other parties who signed the agreement. The board will exist until full and final reclamation of the mine. As per the environmental agreement, the board is funded by Diavik.

Socioeconomic impacts of IBAs and the Tłıchǫ Agreement on Tłıchǫ communities

A review of the scattered evidence on the effect of IBAs on communities, points to the important role played by these agreements in providing socioeconomic benefits, and at times vital services, to some of the Indigenous communities (Meerveld 2016). This is especially important when placed within the context of the rise of the neoliberal state in Canada. In many communities, many of the social support services are provided by mining companies under IBAs or general corporate social responsibility provisions.

Generally, there is a lack of data on the effects of IBAs and CLCAs on the lives of Aboriginal signatory communities. However, a 2014 study by Mehaffey Consulting demonstrated that the Tłıchǫ Agreement has led to improvements in the Tłıchǫ economy and has had a significant impact on the GDP of the Northwest Territories (Tłıchǫ Government 2014). Other positive impacts include increases in business opportunities and mining-related ("spin-off") spending in the region. The study also found that the Tłıchǫ are among the most self-sufficient governments in the North, generating over 75 percent of their own revenues.

In the eight-year period evaluated (2005–2013), the Tłıchǫ Government and Tlichǫ Investment Corporation companies contributed more than CAD450 million to the GDP of the NWT. The Tłı̨chǫ Investment Corporation (wholly owned by the Tłı̨chǫ Government) has a diverse portfolio of firms providing goods and services to the mines. These include Tli Cho Logistics, Tli Cho/Orica Mining Services, Tłı̨chǫ Domco and Air Tindi/Tłı̨chǫ Air.

The Tłı̨chǫ Agreement and IBAs with different mining companies have not only provided a financial windfall for the Tłı̨chǫ Government, but Tłı̨chǫ citizens are also reaping some of the financial benefits of the agreement. According to available data, in 2018, almost half (48.6 percent) of Tłı̨chǫ families earned over CAD75,000 a year compared to 27.7 percent in 2003 (two years before the Tłı̨chǫ Agreement was signed) (GNWT 2020b). On the other hand, the percentage of Tłı̨chǫ families making less than CAD30,000 decreased marginally from 30.8 percent in 2003 to 27 percent in 2018 (GNWT 2020b). The percentage of Tłı̨chǫ families earning between CAD30,000 and CAD75,000 declined from 41.5 percent to 24.3 percent in the same period (GNWT 2020b). These figures show a marked improvement in the earnings of many Tłı̨chǫ families.

Significantly, the unemployment rate has declined and labor market participation rates in Tłı̨chǫ communities have witnessed a marked increase as more people gain employment in the mines. In 1999, Tłı̨chǫ communities had a combined unemployment rate of 43 percent compared to 28.1 percent in 2019. The participation rate also increased from 55.6 percent in 1996 to 60.6 percent in 2019. It is important to note that these figures have fluctuated between the intervening years (GNWT 2020c).

There has also been generous support for social, cultural, educational and other community activities in Tłı̨chǫ communities (Rio Tinto n.d.; GNWT 2018). Scholarships financed from IBA transfers, skills training and other educational programs are increasing community and individual capacity. In 2005, a regional trades and technology training center was established to train local people in essential job-related skills, particularly those applicable in the mining industry. There has also been support for Tłı̨chǫ cultural activities and for institutions providing services to people with addictions and other social problems.

While data on the negative impacts of mining on Tłı̨chǫ communities are scarce, evidence from amalgamated data for small communities in the NWT points to some potentially significant negative impacts of mines on these communities. Increases in discretionary income have led to increased crime, substance abuse and gambling among some in the communities, including Tłı̨chǫ communities (Davison and Hawe 2012; GNWT, 2015). It has also been observed that there is less supervision for young people, as material goods and gifts are replacing appropriate parenting. Also, the inequitable distribution of resources is creating haves and have-nots in communities (Davison and Hawe 2012; GNWT, 2015). While it is difficult to ascertain whether these changes are all solely the result of the booming mining sector alone, it is evident that mining is a contributing factor.

In comparing IBAs in Canada and Australia, O'Faircheallaigh (2020) notes that some Indigenous groups are reaping substantial economic benefits from IBAs and are achieving a significant role in environmental management of developments on their territories. It can be concluded that the Tłı̨chǫ IBAs and the Tłı̨chǫ Comprehensive Agreement have provided substantial economic benefits to Tłı̨chǫ communities. These agreements have also given the

Tłı̨chǫ important roles in the environmental governance structures overseeing mines on or adjacent to their territories.

Hence, it is evident that IBAs and CLCAs are relatable, and both work in tandem to structure the governance mechanism between Indigenous peoples and industry in the Mackenzie Valley. While IBAs are supra-regulatory, they impact the regulatory process in important ways. For example, in the NWT, when a major resource development project is going through the approval process, the territorial government asks the proponent to provide an outline of its programs aimed at offering socioeconomic benefits to northerners and monitoring some of the effects of their project on potentially impacted communities. Since IBAs are confidential agreements, the details of the agreement are not disclosed, but their mere presence is considered vital for project approval.

Prior to the CLCAs, industry and government were the dominant players in the mining industry. The land claims agreements and the requirements to consult with Aboriginals have created many avenues of interaction and increased the frequency of interaction between Indigenous people and the mining industry. Indigenous people are also involved in the governance framework for the mining industry. These interactions have become necessary for the goals of the IBAs and the CLCAs to be met. Impact benefit agreements and CLCAs have not only impacted the socioeconomic aspects of Tłı̨chǫ communities but have also resulted in governance changes in the region. These changes are discussed next.

Tłı̨chǫ Agreement, impact benefit agreements and the emergence of a new governance structure

Abbott and Snidal (2009, p. 52) outline the conceptual model of the "governance triangle" to help represent the groupings of actors and interests in governance schemes that create rules and processes for participating actors. There are three major groupings of actors: "firm," which is composed of individual companies as well as industry associations and other groupings of companies; "NGO," which is a broad category of civil society groups, international non-governmental organizations, academic researchers, activist investors, and individuals; and "state," which includes both individual governments and supranational groupings of governments (e.g., the European Union, the United Nations). The triangle typically includes many governance schemes, with location of each regulatory initiative based on the approximate amount of influence ("governance shares") that each actor exerts over it.

Historically, in the NWT, before the modern treaties were signed, the governance structure in Canada's extractive industry was dominated by the state and industry. However, due to the negative impacts of extractive activities on the livelihood and culture of Aboriginals, the dominance of these actors has been questioned by Aboriginals (Jamasmie 2014). In the Mackenzie Valley, the duty to consult, modern treaties, a more inclusive EA regime and IBAs (Harvey and Bice 2014) have facilitated the emergence of a flexible, innovative and participative governance regime for the interaction between Aboriginal communities and industry.

This governance innovation mirrors the shift from government to governance (a shift from hierarchical representational government by institutions under majority rule to more networked egalitarian stakeholder relations based on consensus (Jones 1998; Braithwaite and Levi 2003). These governance arrangements are not character-ized by the exercise of hierarchical governmental authority but by informal networks that focus on partnerships and networks. In many ways it can be argued that this new governance model is a tool facilitated by the Canadian federal government through a wide range of measures (both legal and administrative) aimed at downloading to industry some of its responsibilities to consult with and accommodate Aboriginal concerns related to mining activities. A counter argument is that it is important for industry to agree to the consultation because they are on the ground and understand the context better. Nevertheless, the fact that government does not play an active role in the negotiation of IBAs somehow tips the balance in favor of industry. The promise of lucrative confidential financial agreements at the end of a long and tiring approval and regulatory process increases the possibility of the scales being turned in favor of industry and against Aboriginal communities in both consultation and during EA.

From the different forms of interaction and governance schemes between the Tłįchǫ Government, the mining industry and other stakeholder organizations, it is evident that the interrelated processes of the federal duty to consult, the independently-run EA process, modern treaties and the locally negotiated IBAs have led to the creation of component parts on different levels of an embedded governance structure for the interaction between industry and Indigenous people. Hence one can identify an interrelated, nested and multiscale governance structure emerging from four distinct governance features (consultation, Tłįchǫ Agreement, EA and IBAs) that can be viewed as an effort by governments, the mining industry and Aboriginal communities to realize a workable solution to some of the chal-lenges brought about by mining. While Figure 1.1 (Chapter 1) as an ideal type, captures the typical governance triangle within which many Indigenous communities operate, Figure 6.4 provides a visual representation of the various

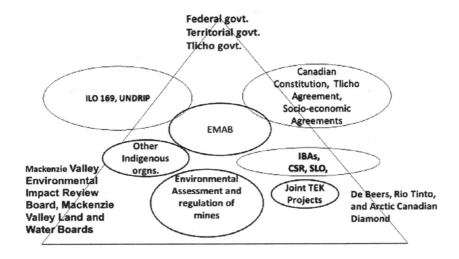

Figure 6.4 Governance Triangle depicting the Tłįchǫ governments interaction with the mines and other stakeholders.

governance schemes within which the Tłı̨chǫ government, mining companies and other stakeholders participate.

Conclusion

For many years, the mining industry gained access to Indigenous lands for the extraction of natural resources without much input from the communities where they operated. However, in northern Canada over the past two decades, comprehensive land claims agreements have provided a robust legal framework to ensure that Indigenous people actively participate in the management of resources on their traditional lands. With the signing of these agreements, Indigenous groups have gained a high degree of political leverage in northern resource governance and are becoming active participants in the regulation of the extraction of resources from their traditional homelands.

Comprehensive land claims agreements offer legal and economic certainty by providing clarity regarding issues such as land ownership and Indigenous rights. This clarity serves to deliver certainty for businesses which, according to some analysts, has driven the Canadian government to negotiate such agreements in resource-rich areas. The agreements have also given communities a strong role in resource management and leverage in negotiating IBAs.

Impact benefit agreements, which in the northern Canada context are legally required, have also empowered Indigenous people to actively participate in the mitigation of the effects of mining on their territories, while also negotiating for more of the benefits accruing from the extraction of resources on their territories.

In this chapter, I have outlined the role of the Tłı̨chǫ Agreement in shaping the relationship between industry and the Tłı̨chǫ. It is evident that the agreement has altered the Tłı̨chǫ people's relationship not only with the state but also their relationship with industry and their role in the regulation of mining activities in the Mackenzie Valley. The Tłı̨chǫ are active participants in the regulation of mining, and environmental initiatives using Tłı̨chǫ traditional knowledge (largely funded by industry under the IBAs) have been undertaken to address some of the adverse impacts of mining operations.

It can also be concluded that IBAs have not only provided substantial economic benefits to Tłı̨chǫ communities, they have also given the Tłı̨chǫ important roles in the environmental management of mines on or adjacent to their territories.

Although the negotiation of impact benefit agreements is a private endeavor between the Tłı̨chǫ Government and industry, IBAs have implications both for the relationship between the Tłı̨chǫ and the mining companies and for the Tłı̨chǫ's relationship with public mining regulatory agencies. The incorporation of some of the funding for ongoing environmental and social monitoring of the project in the IBAs is testimony to the blurred lines between impact benefit agreements and the formal resource management regime.

The negotiation of new treaties and self-government agreements, the signing of IBAs and the duty to consult have led to a reconfiguration of the structures governing the relationship between industry and Aboriginal communities. The governance of these relationships is now being increasingly played at multiple venues and at multiple scales.

In terms of the governance triangle, it is evident that for the Tłı̨chǫ, the relationship between the state and Indigenous people has very important implications for the relationship between the Tłı̨chǫ and industry. This is largely due to the overarching legal framework provided by the CLCAs, which ensure meaningful Tłı̨chǫ participation in resource management on their territories. The same framework provides for the negotiation of IBAs to mitigate the negative impacts of mining and maximize benefits accruing to the communities from mining activities. Hence, it is not far-fetched to argue that in the case of the Tłı̨chǫ, the relationship between the State and the Tłı̨chǫ is the most important angle of the governance triangle. This relationship has provided the Tłı̨chǫ with political leverage in their relationship with industry, thereby slightly leveling the playing field between the two parties.

References

Abbott, K.W. and Snidal, D. (2009) "The governance triangle: regulatory standards institutions and the shadow of the state" in W. Mattli and N. Woods (eds.) *The politics of global regulation*. Princeton, NJ: Princeton University Press, pp. 44–88. https://doi.org/10.1515/9781400830732.44

Braithwaite, V. and Levi, M. (eds.) (2003) *Trust and governance*. New York: Russell Sage Foundation.

Calder v. Attorney General of B.C. (1973) S.C.R. 313.

Davison, C. and Hawe, P. (2012) "All that glitters: diamond mining and Tłı̨chǫ youth in Behchokǫ̀ Northwest Territories," *Arctic*, 65(2), pp. 214–228. Available at: http://pubs.aina.ucalgary.ca/arctic/Arctic65-2-214.pdf

Debeers Group. n.d. "Operations". Available at https://canada.debeersgroup.com/operations/mining/gahcho-kue-mine (Accessed February 2, 2021)

Ehrlich A. (2016) "*Preliminary screening and environmental assessment processes in a nutshell*," Mackenzie Valley Resource Management Act workshop, January 12–13. Yellowknife: Mackenzie Valley Review Board, Land and Water Boards and the Government of the Northwest Territories.

EMAB (2019) *DDMI traditional knowledge panel session #12: options for pit closure* [Online]. Traditional knowledge panel report, September 12–16. Diavik Diamond Mine, NT. Available at: https://www.emab.ca/sites/default/files/tk_panel_session12_oct_12_final_with_appendices.pdf (Accessed February 1, 2021)

Environmental Monitoring Advisory Board, EMAB (n.d.) *EMAB recommendations* [Online]. Available at: https://www.emab.ca/what-we-do/emab-recommendations (Accessed February 1, 2021)

Fidler, C. and Hitch, M. (2007) "Impact and benefit agreements: a contentious issue for environmental and Aboriginal justice," *Environments Journal*, 35(2), pp. 49–69. Available at: https://fnbc.info/resource/impact-and-bene-t-agreements-contentious-issue-environmental-and-aboriginal-justice

Gibson, G., MacDonald, A. and O'Faircheallaigh, C. (2011) "Cultural considerations associated with mining and Indigenous communities" in P. Darling (ed.) *SME mining engineering handbook*, special edition. Denver, CO: Society for Mining, Metallurgy and Exploration, pp. 1797–1816.

GNWT (2018) Socio-economic agreement report for diamond mines operating in the Northwest Territories [Online]. *Department of Industry, Tourism and Investment, ITI.* Available at: https://www.iti.gov.nt.ca/sites/iti/files/socio-economic_agreement_report_-_2018.pdf (Accessed October 18, 2020)

GNWT (2020a) Gross domestic product by industry [Online]. *NWT Bureau of Statistics.* Available at: https://www.statsnwt.ca/economy/gdp/ (Accessed October 16, 2020)

GNWT (2020b) Family income distribution, 2001–2018, Northwest Territories by community and selected geographic aggregation [Online]. *NWT Bureau of Statistics.* Available at: https://www.statsnwt.ca/labour-income/income/index.html (Accessed January 31, 2021)

GNWT (2020c) Community labor force activity, 1986–2019 [Online]. *NWT Bureau of Statistics.* Available at: https://www.statsnwt.ca/labour-income/labour-force-activity/ (Accessed January 31, 2021)

Government of Canada (1998) *Mackenzie* Valley Resource Management Act [Online]. *Justice laws website.* Available at: https://laws-lois.justice.gc.ca/eng/acts/m-0.2/ (Accessed September 30, 2020)

Government of Canada (2005) Tłı̨chǫ Land Claims and Self-Government Act [Online]. *Justice laws website.* Available at: https://laws-lois.justice.gc.ca/eng/acts/T-11.3/FullText. html (Accessed December, 2, 2020)

Government of Canada (2011) Aboriginal consultation and accommodation: updated guidelines for federal officials to fulfill the duty to consult [Online]. *Aboriginal Affairs and Northern Development Canada (AANDC).* Available at: https://www.rcaanc-cirnac.gc.ca/ eng/1100100014664/1609421824729 (Accessed December, 2, 2020)

Government of Canada (2015) Comprehensive claims [Online]. *Crown–Indigenous Relations and Northern Affairs Canada, CIRNAC.* Available at: https://www.rcaanc-cirnac.gc.ca/ eng/1100100030577/1551196153650 (Accessed December, 2, 2020)

Government of Canada (Crown-Indigenous Relations and Northern Affairs Canada) n.d., "Post-1975 Treaties (Modern Treaties) dataset." Available at https://open.canada.ca/data/ en/dataset/be54680b-ea62-46f3-aaa9-7644ed970aef#wb-auto-6. (Accessed March 23, 2021)

Government of Northwest Territories, GNWT (2015) *Communities and diamonds. 2014 annual report of the Government of the Northwest Territories under the Ekati, Diavik and Snap Lake socio-economic agreements* [Online]. Tabled document 227-17(5), tabled on March 12, 2015. Available at: https://www.ntassembly.ca/sites/assembly/files/td227-175.pdf (Accessed December 2, 2020)

Haida Nation v. *British Columbia, Minister of Forests [2004] 3 S.C.R. 511, 2004 SCC 73.*

Harvey, B. and Bice, S. (2014) "Social impact assessment, social development programmes and social licence to operate: tensions and contradictions in intent and practice in the extractive sector". *Impact Assessment Project Appraisal,* 32 (2014), pp. 327–335. Available at https://www.researchgate.net/publication/265340630_Social_licence_to_operate_ and_impact_assessment

Hitch, M. (2006) *Impact and benefit agreements and the political economy of mineral development in Nunavut. Doctoral dissertation.* Waterloo, ON: Department of Geography, University of Waterloo. Available at: https://uwspace.uwaterloo.ca/handle/10012/992 (Accessed December 2, 2020)

Jamasmie, C. (2014) *Canada's oil sands brace for conflict-ridden 2014* [Online]. *mining.com.* Available at: http://www.mining.com/canadas-oil-sands-brace-for-conflict-ridden-2014- 95551/ (Accessed February 2, 2021)

Jones, M. (1998) Restructuring the local state: economic governance or social regulation? *Political Geography,* 17(8), pp. 959–988.

Land Claims Agreements Coalition (n.d.) *What is a modern treaty?* Available at: https://land-claimscoalition.ca/modern-treaty/ (Accessed February 2, 2021)

Levitan, T. (2012) *Impact and benefit agreements in relation to the neoliberal state: the case of diamond mines in the Northwest Territories* [Online]. Master's thesis. Ottawa, ON: Institute of Political Economy, Carleton University. Available at https://curve.carleton.ca/system/files/etd/e90a8a61-01d4-4bf3-a9e4-c11897163e3a/etd_pdf/f4245b1c0db49d9ec7add07ab8c78b63/levitan-impactandbenefitagreementsinrelationtothe.pdf (Accessed February 2, 2021).

Mackenzie Valley Environmental Impact Review Board, MVEIRB (2017) *Report of the resource co-management workshop* [Online]. January 25–26. Hay River Reserve, NWT: Mackenzie Valley Review Board, Mackenzie Valley Land and Water Board and the Government of the Northwest Territories. Available at: https://mvlwb.com/sites/default/files/news/1301/attachments/resource-co-management-workshop-report-hay-river.pdf (Accessed October 15, 2020)

Mackenzie Valley Review Board, MVRB (2016) *Report of environmental assessment and reasons for decision* [Online]. Dominion Diamond Ekati Corp. Jay Project EA1314-01. February 1. Available at: http://reviewboard.ca/upload/project_document/EA1314-01_Report_of_Environmental_Assesment_and_Reasons_for_Decision.PDF (Accessed October 15, 2020).

McKee, C. (2009) *Treaty talks in British Columbia: building a new relationship.* Vancouver, BC: UBC Press.

Meerveld, D. (2016) *Assessing value: a comprehensive study of impact benefit agreements on Indigenous communities of Canada* [Online]. Ottawa, ON: University of Ottawa. Available at: https://ruor.uottawa.ca/bitstream/10393/34816/4/Meerveld%2C%20Drew%2020161.pdf (Accessed October 18, 2020)

Mikisew Cree First Nation v. Canada, Minister of Canadian Heritage (2005) 3 S.C.R. 388, 2005 SCC 69.

O'Faircheallaigh, C. (2020) "Explaining outcomes from negotiated agreements in Australia and Canada," *Resources Policy*, 70, 101922. Available at: https://doi.org/10.1016/j.resourpol.2020.101922

Taku River Tlingit First Nation v. British Columbia [2004] 3 S.C.R. 550, 2004 SCC 74.

Tinto, Rio (n.d.) *Diavik communities* [Online]. Available at: https://www.riotinto.com/can/canada-operations/Diavik/diavik-communities (Accessed January 28, 2021).

Tłįchǫ Government (2014) *A Strategic Framework and Intentions 2013–2017.* Available at: https://www.tlicho.ca/sites/default/files/documents/government/StrategicFramework20142017.pdf (Accessed February 1, 2021)

Tlicho Government and the Government of the Northwest Territories and Canada (2005) *Tłįchǫ land claims and self-government agreement* [Online]. Available at: https://laws-lois.justice.gc.ca/eng/annualstatutes/2005_1/page-1.html (Accessed February 2, 2021)

Tłįchǫ Research and Training Institute (2018) *Traditional knowledge study for the Diavik soil and lichen sampling study [Online].* Available at: https://research.tlicho.ca/research/traditional-knowledge-study-diavik-soil-and-lichen-sampling-study (Accessed February 1, 2021)

Wek'èezhìi Land and Water Board (WLWB) (May 29 2017) Land Use Permit (LUP) W2013D0007, Dominion Diamond Ekati Corporation (DDEC), Condition #58. Available at http://registry.mvlwb.ca/Documents/W2013D0007/Ekati%20Jay%20Project%20- %20Land%20Use%20Permit%20-%20May%2029_17.pdf

7 The shifting state

Rolling over Indigenous rights in Ontario, Canada

Gabrielle A. Slowey

Introduction

The provincial north is an integral part of Canada's northern landscape. Its geography and history are rich in stories and experiences of natural resource extraction, from uranium to forestry, from oil and gas to gold, diamonds, nickel and ore. Indeed, the history of the Canadian north reflects the general history of Canada in general as a staples producer and exporter (Innis, 1930). It also reflects the history of Indigenous and non-Indigenous relations in particular, which is largely a story of non-Indigenous peoples seeking access to and ownership of Indigenous lands and the resources contained in them. Historically, a myriad of reasons has been offered for the taking of Indigenous lands, including white settlement, but none has been or remains so compelling as the desire for unfettered access to resource development. To that end, important progress has been made to ensure that First Nations treaty rights are respected and protected to better ensure that they participate in and benefit from development occurring on or near their traditional territory. Today, there are new norms, processes and procedures (such as the duty to consult and the increasing use of environmental assessments) that the courts have essentially mandated. These processes are designed to secure First Nation agreement for development projects, and government and industry must engage in them if increased development activity is to occur.

However, what happens when there is a global pandemic? What happens to the processes put in place designed to respect and honor Indigenous–state relations? As government finances run dry trying to keep the economy afloat, there is a real need of government to fill its coffers. As the case of Ontario shows, the whittling away of these requirements ensures unfettered access to Indigenous lands. While First Nations rights in the context of resource development have arguably increased, this chapter demonstrates how quickly and easily those advances can change. Looking at Bill 197, officially the *COVID-19 Economic Recovery Act* (2020) in Ontario, this chapter shows how the state is once again using mining to solve its economic problems. Absolute power over natural resources is retained by the province, and its role as promoter of resource development ultimately overrides its role as protector of Indigenous lands and rights.

DOI: 10.4324/9781003131274-7

Ontario in the Treaty context

Canada is a country rich in natural resources. Indeed, as the famous professor of political economy at the University of Toronto Harold Innis first observed (1930)—and others since have debated (Watkins, 2007; Mills and Sweeney, 2013; McNally, 1981)—Canadian history is the history of "staples" (natural resources such as fur, timber, fish and minerals production and export). However, these resources are not evenly distributed across the country and are more commonly located across the Canadian "North" (both territorial and provincial). For instance, in Ontario in 1903 silver deposits were found in Cobalt in northern Ontario, and mining has since been a key economic driver for the region. With the Klondike Gold Rush in the far-off Yukon Territory coming to a close, governments and prospectors alike were keen to find the next hot commodity. The discovery of a silver vein kicked off a rush to northern Ontario, where the town of Cobalt became the focus of mining in 1905–1906. Eventually gold rushes in Kirkland Lake and Timmins would become more appealing and more profitable (MikeyMike426, 2020). Indeed, northern Ontario has been a leader in mining for over a century, responsible for one-third of Canada's total mined metal production (Burkhardt, Rosenbluth and Boan, 2017). The economic opportunity for new mines in Ontario is once again the focus for the Government of Ontario as global demand for minerals continues to grow.

At the same time that gold was being discovered across northern Ontario, the governments of Canada and Ontario appointed three representatives to "negotiate" Treaty 9 to open up northern Ontario for development and settlement (Long, 2010). These treaty commissioners were dispatched to the region to host meetings with local Indigenous peoples, and present to them the terms of the treaty. They arrived with a pre-written document that followed the format of previously negotiated treaties, with some provisions like hunting and trapping taken directly from Treaty 8. Treaty 9 (also known as the James Bay Treaty and not to be confused with the James Bay and Northern Quebec Agreement signed in 1975) was finalized in 1905–1906. Subsequent *adhesions* in 1929–1930 included those northern First Nations that had been left out of the original signing (Long, 2010).

Despite having a treaty, First Nations have had very few, if any, opportunities to influence development. Instead, these Indigenous peoples and their interests have been largely ignored in mining and forestry development in northern Ontario. In fact, what is important in, and unique to, Treaty 9 is the veto power retained by the province of Ontario. Commonly referred to as the "take-up clause," the terms of the written treaty permit the province of Ontario to reclaim any land set aside for Treaty 9 for its own purposes, from fishing to mining to economic development. Eleven numbered/historic treaties were negotiated across parts of Canada in 1871–1921, but unlike in all other Canadian historic treaties, the province was both a signatory to Treaty 9 and also made its own rights explicit. It is important to point out that for the First Nations, what was promised orally (because the chiefs of the day were unable to read the legal document at the time so they relied on the spoken promises) did not reflect what was ultimately found in the written text of the treaty. For instance, according to legal counsel Murray Klippenstein, the negotiators deliberately omitted any mention of the take-up clause in their discussions, which left First Nations duped by verbal promises that they would be able to keep their traditional lands for hunting, fishing and trapping (Wiens, 2014). Hence, Treaty 9 First Nations challenge the validity of the written text as contradicting the promises made orally at the time of the signing, which raises important questions around its interpretation and application (Klippenstein and Quick, 2021). Regardless, the fact that this historic treaty persists today means that Treaty 9 First Nations are bound to the terms of an anachronistic, paternalistic and colonial document unlike those First Nations in other parts of Canada that have more recently negotiated their own, modern treaties.

In Canada, of over 630 First Nations communities representing over fifty nations and fifty language groups, only 26 have settled modern-day treaty land claims negotiated after 1975. In addition, there are 25 self-government agreements across Canada involving forty-three Indigenous communities (of which the Tlichǫ are one, as featured in the chapter by Sam-Aggrey in this volume) (Nunavut governance is achieved through public governance) (Government of Canada, 2020a). The rest of the First Nations across Canada are governed under the *Indian Act* (1876)—a remnant of colonial legislation that continues to define the paternalistic relationship between the state and most First Nations. Those First Nations with a modern agreement are self-determining and are no longer governed under the direct control or jurisdiction of the federal government but interact with the provinces or territories and the federal government on a government-to-government basis.

To elaborate, Canada is a federal country where power is divided between two separate orders of government: the central or federal government and the provinces and territories. The territories are technically creatures of the federal government but because of devolution in Yukon (2003) and the Northwest Territories (2014), power has been transferred to these regions which now function more practically as proto-provinces. Under the terms of the *Constitution Act* (1867), responsibility for "Indians and lands reserved for them" was assigned to the federal government, essentially making Indigenous peoples wards of the state. However, the provinces were given exclusive jurisdiction and important powers over natural resources, education, health and welfare. What this means today is that the provinces have a great deal of power over resource development. It also means that those First Nations that do not have a modern treaty or self-government agreement continue to be administered directly by the federal department of Crown–Indigenous relations and Northern Affairs Canada (CINAC) and Indigenous Services Canada (ISC). In the late 1960s, the federal government attempted to eliminate the First Nations "status" as wards of the state which would have had the effect of ending federal responsibility, closing federal offices and transferring the responsibility for First Nations peoples over to the provinces, but this proposal failed (see the *White Paper* formally known as the *Statement of Government on Indian Policy*, 1969). In addition, most of the remaining 600 First Nations either have a historic treaty relationship—a numbered treaty negotiated in 1871–1921 or a peace and friendship that predates Canadian confederation—or no existing treaty relationship with the federal government (primarily the case in British Columbia). In the absence of any new agreements, these communities remain constrained by historic and ongoing state domination.

The constraints and limitations imposed on First Nations by historic treaties puts those First Nations at a disadvantage, especially when compared to their modern treaty counterparts, because the historic relationship continues to be one of control and domination rather than self-determination. Most modern claims were generated by the need to gain clear legal authority to proceed with mega-resource projects in areas where no treaties previously existed (Slowey, 2021). To that end, the First Nations that have negotiated new agreements under the federal land claims policy have been able to include recent aspirations, advocate for new institutions, such as development corporations and co-management boards, and ensure that their visions and needs are met and mediated through the modern agreement. Under the federal land claims policy, claims cannot be negotiated where there are existing treaties like the numbered treaties, unless, as was the case in the Northwest Territories, where the treaty was egregiously flawed. Although most First Nations in Canada operate local community governments and are thus, theoretically, able to make decisions about their communities' priorities and needs, the reality is that most of these same First Nations across Canada continue to be subject to direct monitoring and oversight by the federal government. There is still an institutionalized relationship of dependence that is in large part tied to the financial condition and reliance of First Nations on the state. Hence, there is asymmetry and unevenness in First Nations experiences both within and across Canada when it comes to how much agency they can exert in the context of mining development in their territories. Where First Nations do not have modern treaties there is no level playing field.

Ontario in the Canadian context

Ontario is one of the four largest provinces in Canada in area (alongside Quebec, British Columbia and Alberta, not including the territories), with a population of almost fifteen million people. It boasts large urban centers (Toronto and Ottawa) and popular sports teams (Raptors in basketball, Blue Jays in baseball and Maple Leafs in ice hockey, all based in Toronto). In terms of sheer size, Ontario covers over a million square kilometers, of which northern Ontario encompasses almost ninety percent of the landmass of the province (almost 807,000 km², southern Ontario being only 140,000 km² and eastern Ontario 35,000 km²). In contrast to its size, northern Ontario has only eight percent of the population (Spotton, 2001). The majority of the provincial population lives in southern Ontario, with over six million people representing almost half of Ontario's total population living in the Greater Toronto Area (GTA) alone.

While Indigenous peoples make up a small proportion of the total Ontario population, there are more Indigenous peoples in Ontario than in any other province (Spotton, 2001). Twenty-three percent of all Indigenous peoples in Canada live in Ontario, with 133 communities located across the province. Five of the twenty largest First Nations in Canada are located in southern Ontario, with Mohawks of Six Nations being the largest (Government of Canada, 2020b). Of those 133 First Nations communities, over thirty First Nations in northern Ontario are considered remote, accessible only by air year-round or ice road. Treaty 9 covers two-thirds of northern Ontario, encompassing an area of 332,000 km², is made up of 38 First Nation communities and has a population of only 17,000 people (Statistics Canada, 2019). Despite the small population, Treaty 9 First Nation communities have a higher percentage of youth, as well as youth out-migration, lower labor force participation and employment rates, lower educational levels and lower income levels than the regional and provincial average (Spotton, 2001).

Ontario may be portrayed as a leader in Canada, but it has not led the promotion and protection of Indigenous rights. The province has historically claimed that it possesses jurisdiction over all of northern Ontario by virtue of Treaty 9 (and Treaty 3, covering the northwestern part of the province), which included the surrender of Aboriginal title to land (Loutit, n.d.). Given this understanding of the treaty, this chapter shows how the governance triangle is not a triangle at all but rather a set of parallel lines, the first set of lines linking government and industry, while the second set of lines links industry and First Nations. Where the triangle hangs off-balance is in the relationship between the province and First Nations. In spite of Supreme Court rulings in Canada that require provinces to consult First Nations on developments that may impact their treaty rights, as Bill 197 demonstrates, the province continues to introduce legislation that effectively reduces regulatory requirements, essentially eschews the duty to consult and makes it easier for industry to access and ultimately develop natural resources.

To clarify, the duty to consult is a constitutional obligation that the courts have found exists in section 35 (1) of the *Canadian Constitution Act* (1982). It is a legal obligation of the Crown (be that federal or provincial) to consult Aboriginal peoples over actions or decisions that may affect their Aboriginal or treaty rights (see Newman, 2009). While the source of the duty to consult lies with the honor of the Crown, the broad purpose of the duty is a process of fair dealing and reconciliation between the Crown and Indigenous peoples. The reality, however, is that the Crown can and typically does delegate the responsibility to gather and share information about a project with First Nations to the industry proponent. This may in part explain the rise, popularity and prevalence of impact benefit agreements, which industries can use in permitting processes to demonstrate First Nations community support for a project. Thus, through a series of rulings starting as far back as *Guerin v. the Crown* in 1984 where the court ruled that the federal government had violated its fiduciary duty by failing to consult with a First Nation, followed by Sparrow in 1990 and the development of the infringement test (the "Sparrow test" sets out a list of criteria that determines whether a right is existing and, if so, how a government may be justified to infringe upon it) and culminating in the Supreme Court trilogy of cases on the duty to consult in 2004, *Haida Nation v. British Columbia (Minister of Forests), Taku River Tlingit v. British Columbia (Project Assessment Director)* and *Mikisew Cree First Nation v. Canada (Minister of Canadian Heritage)* the Supreme Court of Canada has set out a framework of minimal legal requirements. Based on this case law, provinces have been able to draft and develop their own consultation policies which vary greatly across the country from highly prescriptive in some jurisdictions, as in Alberta, to almost non-existent in others, as in New Brunswick.

Ontario is not the only jurisdiction where the focus on resource development during the time of COVID is taking place (Bernauer and Slowey, 2020). From the Yukon to northern Alberta, resource extraction remains a key economic driver (Slowey and Stefanick, 2015). What links these cases is that where there are no modern treaties, the lack of Indigenous rights is amplified, as the development of land remains under the purview of government policies and priorities. First Nations interests that are supposed to be protected by the state can be and often are ignored under the auspices of what government claims serves the greater public, read non-Indigenous, interest. Mining represents one of those "greater interests" that compel governments to act in ways that contravene and contradict their duty to protect and preserve Indigenous rights. This reflects the governments' role as the central actors in the governance triangle of natural resources. However, First Nations are integral to that governance triangle, despite government actions that would suggest the contrary. As Eabametoong Chief Harvey Yesno explains,

> We see ourselves as the key to development in the north. We are the ones that hold the key for certainty, both to government and to industry, who has to raise the money in the stock market. We're one of the legs in the three-legged stool for four-legged stool.
>
> (Garrick, 2020)

While it is relatively common to view conflicts around mining as conflicts between First Nations and industry, the state is the central actor in issuing permits and shaping the conditions under which mining takes place. As the Australian political scientist Cathy Howlett writes:

> States play a key role in the definition and control of resources. They establish property rights, enforce commercial contracts, and regulate the behaviour of the private sector in areas such as company and environmental law. States define the terms on which resources will be accessed, produced, transported, and marketed. In short, states shape the institutional framework within which resource development occurs, and as such are a major determinant of the constraints and opportunities faced by the various actors involved in resource development.
>
> (Howlett, 2010, p. 109)

In recent decades, the Ontario government has been restructured in ways that favor mining corporations. For instance in 1996, the Conservative government issued a series of budget cuts that ultimately limited the capacity of the Ministry of the Environment, the Ministry of Natural Resources and the Ministry of Northern Development, Mines and Forestry to carry out amendments to the Mining Act passed in 1990 that had been aimed at requiring mine proposals to be accompanied by closure and rehabilitation plans to curb the effects of mine abandonment. And as Bill 197 demonstrates, new legislation further opens the gates for development. To demonstrate more clearly how this occurs, I will now turn to explore the case of the "Ring of Fire" in northern Ontario.

"Ring of Fire," northern Ontario: then

The "Ring of Fire" is an area of northern Ontario that has one of the richest deposits in the world of chromite, a key ingredient in the production of steel

(Hjartarson et al., 2014). The area also has other resources—deposits of nickel, copper and platinum—in abundance. Colloquially referred to as the "Ring of Fire" (now rebranded as Wawangajing), the deposit in northern Ontario is among the largest intact wetlands on the planet, 500 kilometers northeast of Thunder Bay in the James Bay Lowlands. The region is more remote by some measures than many parts of the Arctic, and the James Bay Lowlands at the southern tip of Hudson Bay are "at their most phenomenal in spring when they are overwhelmed with bird life. Feds and untried muskeg full of sphagnum moss alternate to form a mottled pattern of string bog." (Gamble, 2017).

The mineral deposit is roughly 330 km from the nearest road or rail line. In 2014, the Ontario Chamber of Commerce released a study on the proposed Ring of Fire development, which covers 5,000 square kilometers, predicting that it would create 5,000 new jobs and generate more than $25 billion in economic activity across numerous sectors in Ontario (McGee and Gray, 2019). However, the development would also affect at least nine First Nations in the immediate vicinity. The Matawa First Nations is a tribal council of ten northern Ontario First Nations which also forms part of the larger political organization, Nishnawbe Aski Nation (NAN) (formerly the Grand Council of Treaty 9) which represents Treaty 3 and Treaty 9 First Nations in northern Ontario. The communities potentially affected by the project include Marten Falls First Nation, Webequie First Nation and Neskantaga First Nation—and others on the edge are Nibinamik First Nation, Aroland First Nation, Long Lake 58 First Nation, Ginnogaming First Nation, Eabametoong First Nation, Miskheegogamang First Nation and Constance Lake First Nation. Five of these communities are not yet accessible by road.

According to an internal briefing note from the federal Department of Aboriginal Affairs,

> First Nations in the Ring of Fire are some of the most socioeconomically disadvantaged communities in all of Canada (plagued by) chronic housing shortages, low education outcomes and lack of access to clean drinking water.
> (McKie, 2013)

All of this jeopardizes the ability of local First Nations to benefit from the potential opportunities associated with the Ring of Fire development. For example, Neskantaga First Nation has been under a "boil water advisory" since 1995, longer than any other First Nation in Canada (Stefanovich, 2020a). In fall 2020, when the community was evacuated to Thunder Bay due to a suspected contamination in the water treatment plant (the water was covered in an oily sheen), Neskantaga Chief Chris Moonias explained in an interview with the Canadian Broadcasting Company how his family was drinking bottled water in the hotel because they were fearful and unaccustomed to drinking water that flowed from the tap (Stefanovich, 2020b). Hence, in principle, the communities agree that the First Nations would benefit from increased roads, increased access to broadband and so forth.

News of the Ring of Fire project first appeared in the provincial budget in 2010 (White, 2020), and shortly thereafter, in May 2011, Matawa chiefs and their communities called for a Joint Environmental Assessment Review Panel.

On October 13, 2011, the Canadian Environmental Assessment Agency (CEAA) proceeded with a Comprehensive Study under Environmental Assessment that favored the mining industry and did not invite local First Nations to participate (Fraser and Rasevych, 2011). Eight days later, on October 21, 2011, Matawa First Nations put out a media advisory, withdrawing its support for the Ring of Fire development unless the federal government agreed to a joint review panel environmental assessment process that would allow First Nations communities in the area to have a voice (Smith, 2011). The Matawa chiefs also announced that, from then on, they would live by the oral treaty because they objected to the "take-up" clause in Treaty 9. So Matawa communities asserted their inherent Aboriginal and treaty rights to refuse impositions created by others. They sought to ensure that any development that takes place on their traditional homelands occurs only with their free, prior and informed consent (commonly referred to as FPIC, as outlined in the United Nations Declaration on the Rights of Indigenous Peoples UNDRIP (see Papillon and Rodon, 2017).

Despite proposed development in the Ring of Fire, major challenges include a lack of access to the remote region, environmental issues and infrastructure such as roads, railroads, electricity and First Nations land rights. For his part, in 2013, the federal minister responsible for the Ring of Fire at that time, Tony Clement, clarified that the only way any extraction project could work was if First Nations people in the region were included as partners. During his time as minister, Clement promised that the federal government would consult with local communities and develop plans to allow them to participate in the economic activity that this project was going to generate. In the end, however, funds were not committed to that purpose.

Then, on June 13, 2013, Cliffs Natural Resources, which owned the deposit announced it was indefinitely suspending its $3.3 billion project, pending further negotiations between First Nations and the province. At that time, Cliffs claimed that the provincial government had not consulted with First Nations, nor was it developing the necessary infrastructure—such as roads and airstrips—required to extract the resource. Shortly afterwards, the company shut down its operations. Still another company, Northern Superior Resources Inc., filed a lawsuit in November 2013 against the Government of Ontario, also charging it with failure to consult with First Nations even though First Nations groups had announced that they were ready and willing to enter bilateral negotiations (CBC News, 2013).

In March 2014, a Regional Framework Agreement was signed between the Matawa First Nations and the Ontario government, giving First Nations in the area a stake in how the Ring of Fire would be developed. The agreement outlined how the nine First Nations that comprise the Matawa First Nations were to work with the province on the environmental assessment process and monitoring. The agreement also addressed matters pertaining to resource revenue sharing and developing regional and community infrastructure. This was an important agreement because economic, social and governance issues (and their interrelationship) were given equal consideration. The framework called for mutual respect, understanding, participation and accountability. Webequie First Nation chief Cornelius Wabasse commented on the importance of a framework agreement:

We just want a proper consultation … and also to work with government side-by-side on how we're going to alleviate some of these issues that will arise from the development in our area.

(Smith, 2011)

Subsequently, in May 2014, the Ontario government announced that it was recommitted to spending $1 billion to build a highway to the province's remote northern Ring of Fire region, with or without a federal government commitment to spending. The potential for resource development to transform northern Ontario was seen as necessary by local First Nations and their leaders, given the poor financial condition of many communities. Indeed, there was a great sense of optimism and hope attached to the development of the Ring of Fire. As Brian Davey, former executive director of the Nishnawbe Aski Development Fund, put it: "Resource development, if it is done right and respects the land, can be a contributing factor in achieving for our communities all four elements (healthy, happy, loving and fulfilled)" (Davey, 2014). For many years the communities of northern Ontario have been repeatedly featured in national news headlines with annual floods, evacuations and calls for relocation. Given the lack of industry in the region, there is high unemployment and, consequently, state dependence. Hence, the desire to benefit from resource development is significant as a vehicle to transform not only the material condition of the communities but also the mentality of the residents. As George Hunter, former chief of Kashechewan, another northern Ontario community, has said,

It's the same problem in Kashechewan. They've got a welfare mentality. The province takes in $400 million a year from licenses. If we just had some of that money, we could look after ourselves.

(Strauss, 2005)

Indeed, the idea of mining as a potential panacea for First Nations poverty continues to inspire support for mineral development projects among First Nations and provincial governments alike.

"Ring of Fire," northern Ontario: now

In 2013 the Ring of Fire development stalled when Cliffs ultimately sold off its interest to Canadian-owned Noront in 2015. Noront now owns 85 percent of the mining claims in the James Bay Lowlands area and has been working toward the completion of the permitting and approval process. In 2019, it was projected that the Eagle's Nest nickel, copper and platinum group mine would be operational by 2024. However, the topic of the Ring of Fire development was made front and center more noticeably when, while on the campaign trail in 2018, then Conservative leader candidate Doug Ford tweeted "If I have to hop on a bulldozer myself, we're going to start building roads to the Ring of Fire" (McGee and Gray, 2019). Promising to be a government of "doers" unlike the previous Liberal government that he labeled a government of "talkers," Ford made it clear that the building of roads to the rich mineral deposit would be a top priority for his government. That is, Ford promised to fulfill a promise made by the previous Liberal government of Kathleen Wynne, which in 2017 had promised provincial support to build an access road. Ford went as far as calling the access road the "Corridor to Prosperity" because it would not only offer mining companies access to the region but would also open up access via road to half a dozen remote First Nation communities that currently have no road access (Kenney, 2019).

Upon taking office in 2018, the Ford government decided to embark on a new process and scrapped the 2014 Regional Framework Agreement proceedings with Matawa First Nations. Apparently, the province wanted a process that would expedite access and ultimately resource production (Northern Ontario Business, 2019). To that end, Noront Resources went directly to local First Nations to negotiate a Memorandum of Understanding with Marten Falls and Aroland First nations, which both became Noront shareholders. Working directly with the local First Nations, Noront promoted the potential benefits of the project as inherent. In the words of a press release:

> For communities like Marten Falls, it is an unprecedented opportunity to transform our socioeconomic future. The youth of Marten Falls look forward to the Ring of Fire as a generational opportunity that can provide training, employment, business prospects, new revenue for social services and many other opportunities – direct and indirect – for the province. Without the Ring of Fire, economic prosperity for our communities will remain a pipe dream.
>
> (Nation Talk, 2019)

Fast forward a few months to 2020 when a global pandemic gripped the country and effectively shut down key economic drivers in the provincial economy. As a result, provincial coffers ran dry in an effort to keep the provincial economy afloat. At the same time, as was the case in most jurisdictions across the country, mining was deemed an essential service: mines could continue to operate and mine workers could continue to travel to and from the mine site to service the mines despite regional and provincial lockdown and stay-at-home orders (Bernauer and Slowey, 2020). To reiterate, at a time when people were told to avoid all non-essential travel, when communities went into full lockdown, with travel bans in place and all in-person, face-to-face meetings canceled, mines and their crews were allowed to continue to operate and to travel relatively unimpeded to and from their homes to the mine sites. In the midst of this pandemic, on July 8, 2020, the Ford government introduced Bill 197: The COVID-19 Economic Recovery Act. This sweeping, omnibus legislation altered twenty pieces of legislation, with amendments to the Building Code Act, City of Toronto Act, Education Act, Payday Loans Act, Marriage Act, Occupational Health and Safety Act, to name a few. However, one particular section modified key parts of the provincial environmental assessment regulations, removing any public appeals as well as the approval of a Provincial Land and Development Facilitator, which "seems purpose-built to fast track development, including low-density sprawl on farmland, forests and wetlands, outside of normal planning processes" (Gray, 2020). The legislation also expanded the use of

Minister's Zoning Orders (MZOs) under the Ontario Planning Act. Originally intended to be used in exceptional cases, the power of MZOs allows a minister to make a ruling on how a piece of land is to be used in the province with no chance of appeal by municipalities, citizens or environmental groups. In 2020, fourteen MZOs were issued for residential/mixed commercial residential projects and another six for industrial/commercial/logistic projects alone (Gray, 2020).

On July 21, 2020, the Bill received its second reading, third reading and royal assent. The speed with which the Bill was passed was very unusual and somewhat alarming, especially for such a major piece of legislation. It is also unusual that the Bill was passed despite no public consultation in the name of economic recovery during this pandemic. Although there is a precedent for the Ford government pushing through legislation without consultation (an Ontario court previously found the Ford government had ignored consultation laws when it canceled the province's cap-and-trade program in 2018), in this case it was clear that the government was using the popular and media preoccupation with COVID to push the legislation through, and heave mining and northern development onto the backs of First Nations (Boisvert, 2019). Although Ontario First Nations have constitutionally protected treaty rights and the courts have mandated consultation in advance of development projects, what the Ford legislation demonstrates is how quickly states can shift and use legislation to undermine established procedures and protections. Conflicts between the state and First Nations that have taken years to resolve can be quickly reversed by the stroke of a pen.

In effect, the legislation reverts Ontario's environmental protection standards back to the 1970s, "when the default was zero assessment and minimal regulatory oversight unless political decision-makers found it expedient" (Bowman, 2020). Under the old rules, the constitutional duty to consult First Nations about development on their traditional lands was triggered by the environmental assessment process. As one lawyer explains, "the new law sets out a new category of 'projects' for which an environmental assessment is 'very much on an Environment Minister's whim'." (CBC News, 2020). Commenting on the new legislation, Executive Director and Counsel at the Canadian Environmental Law Association Theresa McClenagahn said, "For the most part, the proposed changes serve to speed up development at the expense of environmental protection and public participation rights." (Bell, 2020).

Passed without public consultation, Bill 197 was met with alarm and concern among northern First Nations who felt that Ontario was once again undermining treaty rights and shirking its constitutional duties. For their part, Matawa chiefs expressed concern that the province was revamping the environmental assessment process, essentially creating legislation aimed at cutting red tape, including regulatory oversight and monitoring of the environmental impact of natural resource extraction. As Nibinamik First Nation Chief Sheldon Oskineegish put it:

> Our rights cannot be swept under the rug by Crown governments passing legislation designed to clear the way for mining and development on our lands without our consent. It's shameful that Ontario is proceeding in this way and

attempting to use the COVID-19 global pandemic as a smokescreen to ignore their constitutional duties to First Nations. Nibinamik will not stand for such dishonorable action. Any developments or decisions over our Homelands must be made in deep partnership with us.

(Northern Ontario Business, 2020)

In effect, regional chiefs across northern Ontario see Bill 197 as a way for the province to open doors for northern resource development without securing their consent. Although it is true that the courts in Canada have mandated consultation take place, the onus remains on the state and not on companies to ensure that consultation occurs. Corporate behavior has changed somewhat since court rulings (Peerla and Pasternak, 2014), but as Howlett points out, this change is also reflective of a shift in global culture where companies recognize the importance of establishing strong working relationships with local Indigenous peoples (Howlett, 2010, p. 109). The implication here is that the duty to consult and the change in corporate behavior are linked to greater agency exhibited by First Nations. But as the passage of Bill 197 suggests, state governments can still ride roughshod over Indigenous rights, if and when they see fit. Once again, as Howlett (2010) suggests in her review of the Century Mine case in Australia, even with court decisions allowing for greater agency, political actors use legislation to weaken the position of Indigenous peoples (Howlett, 2010, p. 117). For his part, Mushkegowuk Council Grand Chief Jonathan Solomon worries that the legislation takes away the duty to consult with First Nations (Rutherford, 2020). In the end these political decisions and legislative choices made by the state serve to undermine Indigenous rights while at the same time putting the onus of development squarely on Indigenous shoulders. As CEO of Noront Coutts correctly observes, the new streamlining will mean that local First Nations and not the company will be taking over environmental assessment of any roads that will pass through their territory (Rutherford, 2020). Hence it is no wonder why local First Nations refer to Bill 197 as bulldozer legislation, in reference to Premier Ford's previous remarks about ensuring the Ring of Fire development takes place. From their perspective, Bill 197 is designed to ensure that the Ring of Fire development proceeds quickly and relatively unimpeded by processes or procedures designed to protect their interests.

In January 2021, Mushgokewuk chiefs called for a moratorium on any Ring of Fire development, citing potential adverse environmental impacts. They worried, for instance, about potential spills of hazardous materials on the wetlands that could harm the local ecosystems (Baiguzhiyeva, 2021). Citing Bill 197, the Mushgokewuk chiefs were concerned about the lack of consultation and the "fast tracking" of development at the expense of proper planning and consultation concerning the wetlands areas. While the communities may have been in support of the project in principle, they are not in support at any cost. Their reaction to Bill 197 reveals that their support for development is not unconditional and that concern for the land and the environment remains paramount. Opening up the James Bay Lowlands to mining and development will permanently change the landscape of one of the largest intact wetlands left on the planet.

Conclusion

It has been my goal in this chapter to show how the province of Ontario is promoting market solutions to solve its economic crisis. It points to the ways in which the state continues to pursue mining at the expense of protecting Indigenous rights. Using mining as a key driver in its economic recovery strategy, the Ford government is once again turning to the marketplace for solutions. It also shows how quickly a province can pivot to put resource development first and concerns about First Nations rights and even public health, second. In this scenario, the "state has to play a central role, creating a system of laws that protects the interests of private property and deterring challenges/challengers" (Iber, 2018). By passing Bill 197, the Ford government is making it easier for industry by removing important environmental protections and licensing requirements. In this instance, it is clear that the costs of this strategy are primarily being borne by those First Nations on whose territory this extraction is to take place. As Riley Yesno observes, "Advancing resource extraction projects on Indigenous lands during a pandemic is an example of governments and industry using a global pandemic and international health crisis to their advantage." Yesno, who is a member of Eabametoong First Nation, which is affected by the Ring of Fire development, adds:

> This weaponization of the health crisis that is happening in these communities and the weaponization of this terrible thing that's going on is a way to get their desired projects put through.
>
> (Porter, 2020)

Ontario is not the only jurisdiction in Canada to promote and support resource extraction during the pandemic. Mining has been declared an essential service in most of Canada. However, in Canadian regions where Indigenous peoples have modern treaties and new negotiated relationships with the state, they have more say on how resource extraction occurs in their territory. Given that it is unlikely there will be a modern treaty in the Ring of Fire area, coupled with little chance of success with court challenges given the text of the treaty, Treaty 9 First Nations have little recourse to challenge the development aside from political mobilization including the use of blockades and engagement of public opinion. Of course, any concerted widespread action and efforts to coordinate public attention are further hampered by the ongoing pandemic which has hit northern First Nations communities especially hard and continues to prevent communities from coming together.

What is at stake therefore is finding a meaningful way to work with provincial and federal governments in terms of a more comprehensive approach to resource development. For the most part, what is also at stake is development itself. But as has been stated time and time again and is certainly the case in many places across Canada, First Nations are not necessarily opposed to development. Neither, however, are they willing to support it at any cost, since they will bear most of the costs in terms of social ills (increased drug use, abuse, alcoholism, and the like) and environmental degradation. The key is to find a balance between economic development and the protection of lands for future generations (Slowey, 2009).

References

Baiguzhiyeva, D. (2020) *New gold mine brings job opportunities for Mattagami First Nation* [Online]. TimminsToday.com. Available at: https://www.timminstoday.com/local-news/new-gold-mine-brings-job-opportunities-for-mattagami-first-nation-2585503 (Accessed March 10, 2021)

Baiguzhiyeva, D. (2021) *Mushkegowuk chiefs call for a moratorium on Ring of Fire development* [Online]. TimminsToday.com. Available at: https://ca.news.yahoo.com/mushkegowuk-chiefs-call-moratorium-ring-192000117.html (Accessed March 16, 2021)

Bell, A. (2020) Omnibus Bill 197: what you need to know [Blog] *Ontario Nature Blog.* Available at: https://ontarionature.org/omnibus-bill-197-what-you-need-to-know/ (Accessed March 10, 2021)

Bernauer, W. and Slowey, G. (2020) "COVID-19, extractive industries, and Indigenous communities in Canada: notes towards a political economy research agenda," *Extractive Industries and Society: An International Journal,* 7(3), pp. 844–846.

Bill 197. 2020. *Covid-19 Economic Recovery Act.* 1st session, 42nd parliament. Ontario.

Boisvert, N. (2019) *Doug Ford government broke the law when it scrapped cap-and-trade, court rules* [Online]. CBC. Available at: https://www.cbc.ca/news/canada/toronto/ontario-cap-and-trade-ruling-1.5317611 (Accessed March 29, 2021)

Bowman, L. (2020) Ontario passes sweeping changes to environmental assessment [Blog] *Ecojustice Blog.* Available at: https://ecojustice.ca/ontario-proposes-sweeping-changes-to-environmental-assessment/ (Accessed March 17, 2021)

Burkhardt, R., Rosenbluth, P. and Boan, J. (2017) *Mining in Ontario: a deeper look* [Online]. Ontario Nature. Available at: https://ontarionature.org/wp-content/uploads/2017/10/mining-in-ontario-web.pdf (Accessed March 10, 2021)

Chretien, J. (1969). *Statement of the Government of Canada on Indian policy, 1969.* Ottawa: Published under the authority of the Honourable Jean Chretien, Minister of Indian Affairs and Northern Development.

CBC News (2013) *Exploration firm sues Ontario for $110M over mining claims* [Online]. Available at: https://www.cbc.ca/news/canada/thunder-bay/exploration-firm-sues-ontario-for-110m-over-mining-claims-1.2251748 (Accessed March 17, 2021)

CBC News (2020) *Ontario using COVID-19 as a "smokescreen" to trample treaty rights, chiefs say* [Online]. Available at: https://www.cbc.ca/news/canada/thunder-bay/bill-197-first-nations-1.5712623 (Accessed March 29, 2021)

Davey, B. (2014) *Making sure First Nations gain from resource development* [Online]. Republic of Mining. Available at: https://republicofmining.com/2014/09/10/making-sure-first-nations-gain-from-resource-development-by-brian-davey-onotassiniik-magazine-fall-2014/ (Accessed March 17, 2021)

Dylan, A., Smallboy, B. and Lightman, E. (2013) "Saying no to resource development is not an option: economic development in the Moose Cree First Nation," *Journal of Canadian Studies,* 47(1), pp. 59–90.

Everett, E. (2020) *Regional governance change in northern Norway: insights for northern Ontario, Canada.* Master's thesis in Governance and Entrepreneurship in Northern and Indigenous Areas [Online]. UiT The Arctic University of Norway. Available at: https://munin.uit.no/bitstream/handle/10037/18855/thesis.pdf?sequence=2&isAllowed=y (Accessed March 30, 2021)

Fraser, A. and Rasevych, J. (2011) *No joint review panel environmental assessment, no Ring of Fire developments say Matawa chiefs* [Online]. MiningWatch, Canada. Available at: https://miningwatch.ca/sites/default/files/media-advisory-matawa-chiefs-media-conference-dr-10-october-01-2011-2.pdf (Accessed March 10, 2021)

Gamble, J. (2017) *What's at stake in Ontario's Ring of Fire* [Online]. Canadian Geographic. Available at: https://www.canadiangeographic.ca/article/whats-stake-ontarios-ring-fire (Accessed March 29, 2021)

Garrick, R. (2020) *More concerns raised over Bill 197* [Online]. Wawatay News. Available at: https://wawataynews.ca/environment/more-concerns-raised-over-bill-197 (Accessed March 29, 2021)

Government of Canada (2020a) *Self-government* [Online]. Available at: https://www.rcaanc-cirnac.gc.ca/eng/1100100032275/1529354547314 (Accessed March 29, 2021)

Government of Canada (2020b) *Indigenous communities in Ontario* [Online]. Available at: https://www.sac-isc.gc.ca/eng/1603371542837/1603371807037 (Accessed March 29, 2021)

Gray, T. (2020) *You may have never heard of a Minister's Zoning Order and that used to be ok – but not anymore* [Online]. Environmental Defence. Available at: https://environmentaldefence.ca/2020/08/28/may-never-heard-ministers-zoning-order-used-ok-not-anymore/ (Accessed March 29, 2021)

Guerin v The Queen [1984] 2 S.C.R. 335.

Haida Nation v British Columbia (Minister of Forests), [2004] 3 S.C.R. 511.

Hjartarson, J., McGuinty, L. and Boutilier, S. with Majernikova, E. (2014) *Beneath the surface: uncovering the economic potential of Ontario's Ring of Fire* [Online]. Toronto: Ontario Chamber of Commerce. Available at: https://occ.ca/wp-content/uploads/Beneath_the_Surface_web-1.pdf (Accessed March 10, 2021)

Howlett, C. (2010) "Indigenous agency and mineral development: a cautionary note," *Studies in Political Economy: A Socialist Review*, 85(1), pp. 99–123.

Iber, P. (2018) *Worlds apart: how neoliberalism shapes the global economy and limits the power of democracies* [Online]. The New Republic. Available at: https://newrepublic.com/article/147810/worlds-apart-neoliberalism-shapes-global-economy (Accessed March 17, 2021)

Innis, H.A. (1930) *The fur trade in Canada; an introduction to Canadian economic history*. New Haven: Yale University Press.

Kenney, K. (2019) *"Corridor of prosperity" to help Ring of Fire development* [Online]. MyNorthBayNow. Available at: https://www.mynorthbaynow.com/48116/corridor-of-prosperity-to-help-ring-of-fire-development/ (Accessed March 29, 2021)

Klippenstein, M, and Quick, P. (2021) *Discussion paper* [Online]. Treaty 9 diaries: the real agreement between First Nations and the Crown in 1905. Available at: http://www.treaty9diaries.ca/materials-and-documents/discussion-paper/ (Accessed March 29, 2021)

Kornacki, C. (2011) *Matawa, Mushkegowuk agree on oral treaty* [Online]. Wawatay News. Available at: https://issuu.com/wawatay/docs/13oct2011 (Accessed March 17, 2021).

Long, J. (2010) *Treaty 9: making the agreement to share the land in far northern Ontario in 1905*. Kingston-Montreal: McGill-Queen's University Press.

Loutit, S. (n.d.). "The Real Agreement As Orally Agreed To" The James Bay Treaty-Treaty No. 9. [Online]. A presentation by Grand Chief Dr. Stan Loutit Mushkegowuk Council. Available at: https://www.oise.utoronto.ca/deepeningknowledge/UserFiles/File/jamesbaytreaty9_realoralagreement.pdf (Accessed April 19, 2021)

McGee, N. and Gray, J. (2019) "The road to nowhere: claims Ontario's Ring of Fire is worth $60 billion are nonsense," *The Globe and Mail*, October 25 [Online]. Available at: https://www.theglobeandmail.com/business/article-the-road-to-nowhere-why-everything-youve-heard-about-the-ring-of/ (Accessed March 29, 2021)

McKie, D. (2013) Ring of Fire mining may not benefit First Nations as hoped: internal memo from Aboriginal Affairs paints troubling picture [Online]. *CBC News*. Available at: https://www.cbc.ca/news/politics/ring-of-fire-mining-may-not-benefit-first-nations-as-hoped-1.1374849 (Accessed March 10, 2021)

McNally, D. (1981) "Staple theory as commodity fetishism: Marx, Innis and Canadian political economy," *Studies in Political Economy: A Socialist Review*, 6(1) pp. 35–63.

Mills, S. and Sweeney, B. (2013) "Employment Relations in the Neostaples Resource Economy: Impact Benefit Agreements and Aboriginal Governance in Canada's Nickel Mining Industry." *Studies in Political Economy* 91(1) pp. 7–34.

MikeyMike426 (2020) *Cobalt: discovering the mining history of northern Ontario* [Online]. Inside Exploration: Multimedia Due Diligence Database. Available at: https://insidexploration.com/cobalt-a-walk-through-history// (Accessed March 10, 2021)

Mikisew Cree First Nation *v.* Canada (Minister of Canadian Heritage), [2005] 3 S.C.R. 388, 2005 SCC 69.

NationTalk (2019) *Statement from Noront Resources and Marten Falls First Nation on Province of Ontario plan to advance the Ring of Fire* [Online]. Available at: https://www.feathersofhope.ca/statement-from-noront-resources-and-marten-falls-first-nation-on-province-of-ontario-plan-to-advance-the-ring-of-fire/ (Accessed March 17, 2021)

Newman, D.G. (2009) *The duty to consult: new relationships with Aboriginal peoples.* Vancouver: UBC Press and Purich Publishing.

Northern Ontario Business (2019) *Province starts over Ring of Fire consultation process* [Online]. Available at: https://www.northernontariobusiness.com/regional-news/far-north-ring-of-fire/province-starts-over-on-ring-of-fire-consultation-process-1660987 (Accessed March 17, 2021)

Northern Ontario Business (2020) *Matawa chiefs accuse Queen's Park of Far North land grab* [Online]. Available at: https://www.northernontariobusiness.com/regional-news/far-north-ring-of-fire/matawa-chiefs-accuse-queens-park-of-far-north-land-grab-2673335#:~:text=The%20chiefs%20of%20a%20northwestern,the%20North%22%20without%20regional%20consent (Accessed March 17, 2021)

Papillon, M. and Rodon, T. (2017) "Proponent–Indigenous agreements and the implementation of the right to free, prior and informed consent in Canada," *Environmental Impact Assessment Review*, 62, pp. 216–224.

Peerla, D. and Pasternak, S. (2014) *First Nations social contracts: how to contain an aboriginal rebellion: are economic development agreements the best alternative or a containment strategy?* [Online]. CBC News. Available at: https://www.cbc.ca/news/indigenous/first-nations-social-contracts-how-to-contain-an-aboriginal-rebellion-1.2654595 (Accessed March 10, 2021)

Porter, J. (2020) *Ontario ignores its own advice, presses First Nations to consult on Ring of Fire road during COVID-19* [Online]. CBC News. Available at: https://www.cbc.ca/news/canada/thunder-bay/first-nation-consultation-covid-1.5761963 (Accessed March 10, 2021)

R v Sparrow, [1990] 1 S.C.R. 1075.

Ross, I. (2020) *The drift 2020: sharing the mineral wealth* [Online]. Northern Ontario Business. Available at: https://www.northernontariobusiness.com/drift-2020-/sharing-the-mineral-wealth-2127723 (Accessed March 10, 2021)

Rutherford, K. (2020) Mushkegowuk Grand Chief raises concerns about new law that changes environmental assessment process [Online]. CBC News. 13 August. Available at: https://www.cbc.ca/news/canada/sudbury/mushkegowuk-grand-chief-environmental-bill-concerns-1.5684040 (Accessed April 19, 2021)

Slowey, G. (2009) "A fine balance: Aboriginal peoples in the Canadian North and the dilemma of development" in A.M. Timpson (ed.) *First Nations, first thoughts: the impact of Indigenous thought in Canada.* Vancouver: University of British Columbia Press, pp. 229–247.

Slowey, G. (2015) "Game-changer? Resource development and First Nations in Alberta and Ontario" in M. Papillon and A. Juneau (eds.) *Canada: the state of the federation 2013. Aboriginal governance.* Montreal and Kingston: McGill-Queen's University Press, pp. 171–188.

Slowey, G. (2021) *Indigenous self-government in Yukon: looking for ways to pass the torch* [Online]. IRPP Insight (Centre of Excellence on the Canadian Federation, Institute for Research on Public Policy). Available at: https://centre.irpp.org/research-studies/indigenous-self-government-in-yukon-looking-for-ways-to-pass-the-torch/ (Accessed March 29, 2021)

Slowey, G. and Stefanick, L. (2015) "Development at what cost? First Nations, ecological integrity and democracy" in M. Shrivastava and L. Stefanick (eds.) *Alberta Oil and the decline of democracy in Canada.* Athabasca: Athabasca University Press, pp. 195–224.

Smith, J. (2011) *Removing support* [Online]. tbnewswatch.com. Available at: https://www.tbnewswatch.com/amp/local-news/removing-support-389059 (Accessed March 10, 2021)

Spotton, N. (2001) *A profile of Aboriginal peoples in Ontario* [Online]. Ontario Ministry of the Attorney General. Available at: https://www.attorneygeneral.jus.gov.on.ca/inquiries/ipperwash/policy_part/research/pdf/Spotton_Profile-of-Aboriginal-Peoples-in-Ontario.pdf (Accessed March 29, 2021)

Statistics Canada (2019) *Aboriginal population profile, 2016 Census: Treaty 9 – Ontario [Historic treaty area] Ontario* [Online]. Available at: https://www12.statcan.gc.ca/census-recensement/2016/dp-pd/abpopprof/details/page.cfm?Lang=E&Geo1=AB&Code1=2016C1005045&Data=Count&SearchText=Treaty%209%20%2D%20Ontario&SearchType=Begins&B1=All&GeoLevel=PR&GeoCode=2016C1005045&SEX_ID=1&AGE_ID=1&RESGEO_ID=1 (Accessed March 29, 2021)

Stefanovich, O. (2020a) *Members of Neskantaga come home to boil water advisory today* [Online]. CBC News. Available at: https://www.cbc.ca/news/politics/neskantaga-boil-water-advisory-water-operator-1.5841277 (Accessed March 10, 2021)

Stefanovich, O. (2020b) *Stuck in a hotel during a Christmas pandemic, Neskantaga members wait for water crisis to end* [Online]. CBC News. Available at: https://www.cbc.ca/news/politics/stefanovich-neskantaga-water-thunder-bay-evacuation-1.5822885 (Accessed March 10, 2021)

Strauss, J. (2005) "Welcome to Peawanuck, a Reserve that Works" *Globe and Mail* 12 November. Available at: https://www.theglobeandmail.com/news/national/welcome-to-peawanuck/article1130740/ (Accessed April 19, 2021)

Taku River Tlingit First Nation *v.* British Columbia (Project Assessment Director), [2004] 3 S.C.R. 550, 2004 SCC 74

Warick, J. (2021) *Saskatchewan First Nation erects blockade after company enters territory without consent* [Online]. CBC News. Available at: https://www.cbc.ca/news/canada/saskatoon/sask-first-nation-blockade-territory-company-1.5920039 (Accessed March 10, 2021)

Watkins, M. (1977) "The staple theory revisited," *Journal of Canadian Studies*, 12(5), pp. 83–95.

Watkins, M. (2007) "Staples redux," *Studies in Political Economy: A Socialist Review*, 79(1), pp. 213–225.

White, E. (2020) *Ring of Fire was "overhyped," but still "worth the effort" says Noront Resources* [Online]. CBC. Available at: https://www.cbc.ca/news/canada/sudbury/ring-of-fire-10-years-later-1.5479300 (Accessed March 29, 2021)

Wiens, M. (2014) *Tsilhqot'in First Nation ruling means revisiting the James Bay Treaty, says lawyer* [Online]. CBC. Available at: https://www.cbc.ca/news/canada/toronto/tsilhqot-in-first-nation-ruling-means-revisiting-the-james-bay-treaty-says-lawyer-1.2690140 (Accessed March 29, 2021)

8 Emerging governance mechanisms in Norway

A cautionary note from the Antipodes

Catherine Howlett and Rebecca Lawrence

Introduction

The Norwegian state plays a somewhat different role in resource development controversies than the state in settler colonies such as Australia. Settler colonies may be understood as those colonial territories where the settlers came to stay and permanently displaced the Indigenous populations within their acquired territories (Maddison, 2013). Whereas Australia, like other neoliberal Anglo-Saxon states, traditionally sees the state as having a minimal role, Norway is a strong welfare state with an expansive role. It is actively engaged in land use planning, with centralized decision-making powers on resource developments. It therefore came as a surprise to read in 2019 that one of the reindeer herding communities that would be affected by a mega-wind farm development in Finnmark, northern Norway, had signed an impact benefit agreement (IBA) with the company. According to the leader of the reindeer herding district, the community opposed the development, but agreed through the IBA not to oppose the project, and not use legal or other means to challenge the validity of project approvals. Lars Huemer, professor at the Department of Strategy and Entrepreneurship at BI Norwegian School of Management, has commented in the Norwegian media that the agreement, and others of its type, are about companies wanting to respect Indigenous peoples and to do a better job in Indigenous territories (Larsen, 2018). This chapter is written as a salutary note of caution for those in Norway who promote agreement-making mechanisms as effective resource governance tools and as a way to secure respect for Indigenous rights. It urges consideration of the neoliberal logic in which these agreements are embedded, a logic which, while presenting itself as democratic and participatory in reality reinforces existing conditions of injustice and facilitates the further dispossession of Indigenous peoples of their traditional lands.

In Australia, under neoliberalism, the state has been gradually, but consistently, removing itself from its obligations to protect Indigenous peoples from the expansion of private property and resource developments, encouraging private agreement-making as the dominant form of resource governance and downloading the responsibility for deal-making to industry and Indigenous "stakeholders." Canadian political scientists Martin Papillon, Dominique Leydet and Jean Leclair (2020, pp. 1–10) point to similar processes in Canada, where Indigenous participation in decision-making over extractive projects on their lands is channeled through highly regulated impact assessment procedures and the negotiation of IBAs with

DOI: 10.4324/9781003131274-8

project proponents. The focus of negotiations for these projects is often the content of the compensation package Indigenous communities can bargain in exchange for their consent, rather than the impacts of the project or the total real profits of the project for the developer (Papillon and Rodon, 2019, p. 324).

We suggest that the trend to agreement-making so apparent in the settler colonies can be read as a neoliberal one. We acknowledge that Indigenous agency is an important factor in terms of agreement outcomes. Indeed, as Australian political scientist, Cairan O'Faircheallaigh has demonstrated, while legislative frameworks that facilitate agreement-making have a strong impact on negotiation outcomes, they need not be determinative. O'Faircheallaigh (2016, p. 202) demonstrates through a rigorous study of forty-five negotiated agreements in Australia that in spite of legislative and structural inequalities, some Indigenous communities have succeeded in negotiating agreements that are "strongly favourable to Aboriginal interests" thanks to Aboriginal land organizations, the individual agency of Aboriginal communities and learning processes within those collectives (2016, p. 202). We are concerned, however, that there is an overemphasis on Indigenous agency as a critical factor in agreement outcomes, without an acknowledgment of the institutional and market constraints (which we argue are neoliberal in character) that inform this agency. Following scholars Deidre Howard-Wagner, Maria Bargh and Isobel Altamirano-Jiménez (2018, p. 2), we argue that examining agreement-making via a neoliberal lens does not preclude or dismiss Indigenous agency. On the contrary, this chapter seeks to unpack the effects of neoliberalism upon Indigenous agency in agreement-making processes in Australia. It also, to a lesser extent, draws upon some lessons for Norway from the Swedish context, where agreement-making is arguably more advanced. There are salutatory lessons to be gleaned from such an analysis for those Sámi communities who are currently dealing with resource development pressures on their lands and territories in the Artic regions. Initiating such a discussion on the neoliberal logic that informs agreement-making in settler colonies such as Australia will ultimately enhance the agency of those Sámi communities who engage in resource governance decision-making processes in the Arctic region, an area under increasing resource development pressure.

What is neoliberalism?

At its most abstract, it is an assemblage of coercive practices tending always to reinforce existing relations of power founded in control of the economy.

(Sullivan, 2018, p. 201)

Neoliberalism is most commonly understood as enacting an ensemble of economic policies and practices based upon the root principle of affirming the free market. The policies and principles include, but are not limited to: deregulation of industry and capital flow; the end of wealth redistribution as an economic or social political policy; the conversion of every human need or desire into a profitable enterprise; and the increasing dominance of finance capital in the dynamics of the economy, with an ever-increasing intimacy of this finance and corporate capital with the state

(Brown, 2015, pp. 28–29). Similarly, geographers Kim England and Kevin Ward (2007, pp. 3–7) summarize neoliberalism as an economic and political orthodoxy marked by commitments to policies of free trade, privatization, deregulation and welfare state retrenchment. They also contend that it encompasses such issues as the cutting of public expenditure on social services, the elimination of the concept of "public goods" and the restructuring of the welfare state (England and Ward, 2007).

For geographers Nik Heynen, James McCarthy and Scott Prudham (2007, pp. 16–17), the term refers to an economic and political philosophy that questions, and in some versions entirely rejects, government interventions in the market and people's relationships to the economy, and eschews social and collective controls over the behavior and practices of firms, the movement of capital and the regulation of socioeconomic relationships. New Zealand political scientist Maria Bargh (2007) describes neoliberalism as those practices and policies which seek to extend the market mechanism into areas of the community previously organized and governed in other ways. This process involves the entrenching of the central tenets of neoliberalism: free trade and the free mobility of capital, accompanied by a broad reduction in the ambit and role of the state (Bargh, 2007, p. 1). For political economist Damian Cahill (2007), a defining feature of neoliberalism is the transfer of resources from public services to private providers in the name of creating a market for such services, and of fostering choice.

Cahill argued (2007, p. 226) that neoliberalism was the dominant logic of policymaking in most countries. Many commentators decreed that the global financial crisis of 2007 precipitated the demise of neoliberalism as the dominant global ideology (see Howlett et al., 2011). However, we believe that little has changed in the intervening years, and despite those who proclaim that "neoliberalism ate itself" (Denniss, 2018), we contend, following McCormack (2017, p. 7), that the global financial crisis actually strengthened neoliberalism, and that it endures as the contemporary dominant policy logic in most Western nations.

Politically, neoliberalism works to redefine the nature and function of the state (McCormack, 2017, p. 6). Under neoliberalism the state and capital are tightly intertwined, both institutionally and personally (Harvey, 2010, p. 219). While many predicted the demise of the state under neoliberalism, in actuality it activates the state on behalf of the economy, in order to regulate society via the market (Brown, 2015, pp. 62–64). What has occurred under neoliberalism is in fact a repositioning of the state whereby it acts to create conditions in which private corporations can operate more profitably. This often places the state in a contradictory position, simultaneously serving as regulator, investor and development advocate for the private sector (Rigby et al., 2017, p. 19). Finally, under neoliberalism, liberal rights, the foundational cornerstone of traditional liberal states, are extended or withdrawn according to the states' estimate of citizens' capacity to meet their obligations (Sullivan, 2018, p. 202).

Neoliberalism as a form of governance

As political scientist Wendy Brown (2015, p. 122) argues, "governance has become neoliberalism's primary administrative form." Neoliberalism has replaced government—democratic institutions and practices—with governance, new administrative forms that are grounded in business processes, norms and practices. This shift has done real damage to democratic institutions and practices in that it takes contentious resource developments and makes them "more easily 'governable' for both governments and companies by supposedly shifting from adversarial politics to consensual agreements" (Peterson St-Laurent and Billon, 2015, p. 592). This new form of neoliberal governance has

> irrigate[d] every crevice of society: it is through replacing democratic terms of law, participation and justice with idioms of benchmarks, objectives, and buy-ins that governance dismantles democratic life while appearing only to instil it with best practices.
>
> (Brown, 2015, p. 130)

Agreement-making tools arise out of the governance possibilities of neoliberalism. As a technology of governance, these agreements rely on the disengagement of the state and the devolution of responsibilities for negotiations over natural resource developments to private industry. The otherwise public and confrontational debates on the acceptability of natural resource developments are replaced with closed-door meetings with Indigenous leaders, who are encouraged to think of themselves as market actors who "rationally deliberate about alternative courses of action, make choices, and bear responsibility for the consequences of these choices" (Brown, 2015, p. 29). Thus, neoliberal governance strategies create a mirage that those engaging in agreement-making processes do so of their own volition as economically self-reliant individuals who, via their engagement, are deemed to have capacity to negotiate with industry. The analytical focus is thus cast upon "agency" as the secret ingredient that underpins capacity in this milieu. Finally, as a governance model rooted in a self-regulating free market, competition and self-interest that extends to all social realms and concomitantly displaces all other forms of exchange and ethics (McCormack, 2017, p. 6), neoliberal governance transforms cultural, economic and social systems, and this has serious ramifications for Indigenous peoples.

Neoliberalism and Indigenous peoples

Neoliberalism therefore presents significant challenges for Indigenous communities. It is based on universalism, a focus on the individual, a growing intolerance of cultural difference and a limited view of development that is committed to market-based solutions (Altman, 2009, p. 40). Bargh (2007, p. 13) argues, therefore, that neoliberalism is often incompatible with an Indigenous worldview, which is not always easily reconcilable with a market-based, capitalist, neoliberal ethic. Anthropologist William Stanner, with a degree of prescience, suggested that the market and the dreaming may indeed be at variance: "Ours is a market civilisation, theirs not … The Dreaming and The Market are mutually exclusive." (Stanner, 1979, p. 58). Economic historian Karl Polanyi (1957) argued that economies are embedded in social relations and therefore cannot be a separate autonomous sphere vis à vis society as a whole, in direct contradiction to a neoliberal logic. Indeed, within market-dominated societies, it can often become almost impossible for people to imagine that social relations, or human–environment relations for that matter, might possibly be organized differently (Pinkerton and Davis, 2015, p. 305). Yet the reality for many Indigenous communities in Australia is that resource developments offer the only opportunity for economic development and access to the services and opportunities that economic development can encompass. We are not romanticizing or essentializing Indigenous peoples as somehow "naturally" opposed to economic developments on their lands, and we acknowledge the benefits that economic development can potentially bring to Indigenous communities (see Langton and Webster, 2012, O'Faircheallaigh, 2016). What we are saying is that under current neoliberal governance arrangements, nation to nation negotiations between Indigenous nations and states over resource negotiations are not taking place—as they should—because they have been replaced by agreements negotiated between corporate entities and Indigenous peoples. This supplants democratic and accountable government responsibilities toward Indigenous peoples with corporate agreement-making processes that are negotiated behind closed corporate doors, and the proposed benefits from these agreement-making processes are often difficult to assess because of corporate confidentiality arrangements.

Legal scholar David Wishart (2005) holds that in settler colonies such as Australia, Canada and New Zealand, agreement-making appears to be the most commonly adopted strategy in contemporary relations between the settler community and prior occupants. Settler-colonial states aim to replace Indigenous peoples on their land permanently. Unlike other decolonized states, where nominal independence has been obtained from the colonizing force, under settler colonialism, the colonizer never goes home (Howlett and Lawrence, 2019). Settler colonialism is a powerful contemporary force that continues to structure the relationship between Indigenous peoples and the states that have encapsulated them. The Indigenous connection to land is particularly threatening to settler-colonial society, because in Western polities, land is simultaneously a physical commercial resource and a marker of the boundaries of sovereign authority. Thus, in settler societies, such as Australia, a focus on agreement-making as a means to settle issues of land ownership and control implies that Indigenous people "consent" to settler sovereignty. Agreement-making also demands that Indigenous peoples develop governance bodies to enable their participation in these agreements. They become Indigenous bodies that govern themselves in liberal ways, which is another manifestation of their consent to settler sovereignty (Strakosch and Macoun, 2012).

Under neoliberalism, Indigenous peoples, who historically have been excluded from participation in the economies of the colonial states in which they are encapsulated, are encouraged to integrate into the global economy and realize their newly recognized rights to development via the market and self-government. This is a convenient adjustment in settler-colonial relationships that fits well with the reduction of the state and the transfer of administrative responsibilities that characterize neoliberalism (Howard-Wagner, Bargh and Altamirano-Jiménez, 2018, p. 12). Communities are expected to act as individual economic actors, striking deals with powerful corporations, putting up for sale what they have, which is their public support for a project and their free, prior and informed consent. Individual *agency* is foregrounded, structural inequalities and dispossession are backgrounded. Indigenous communities are told, time and time again, that the next resource extraction project could be a game changer for them, bringing "economic development" and lifting the community out of poverty. This narrative accords with the reduced role of the state and the privatization of state assets, functions and services that characterize neoliberalism (Cameron and Levitan, 2014; Howard-Wagner, Bargh and Altamirano-Jiménez, 2018, p. 12).

Agreement-making in Australia

Historically there has been a lack of agreements and treaties between governments and Indigenous peoples in Australia. The new culture of agreement-making that is the focus of this chapter, in Australia at least, is quite novel and has its origins in the Native Title Act 1993 (Howard-Wagner, 2010, p. 102). In 1992, the High Court of Australia handed down the historic *Mabo* decision, which created an opportunity for the Anglo-Australian legal system to recognize interests in land that had existed prior to colonization—native title. The judgment determined that where claimants can show that they have an ongoing connection with land according to their "traditional" laws, and their interest has not been extinguished, then the court *may* declare that their interest be recognized (Galloway, 2020, p. 15). Following the Mabo decision, and amidst heated public debate, the Keating Labor government enacted the Native Title Act 1993 (Cth) (NTA).

State governments and industry consistently argued against the recognition of native title, insisting it should be subordinate to all other titles, and to development. In theory, the intent of the Act was to provide for recognition and protection of native title; its validation and registration; negotiation, mediation and determination of interests; and for compensation (Galloway, 2020, p. 15). There are those, however, who consider that under the NTA, native title has been relegated to an inherently weak and vulnerable title, which can be diminished or extinguished by inconsistent Crown actions (Dillon, 2018). Galloway dismisses the capacity of the NTA to bring justice to Indigenous peoples in Australia after two centuries of dispossession, and argues that the Act "remains constrained by its operation as a tool of the colonising state" (Galloway, 2020, p. 15).

In Australia, the Native Title Act 1993 (and particularly the amendments to the Act enshrined in legislation in 1998, The Native Title Amendments Act 1998) sits firmly at the center of agreement-making. Agreements are explicitly provided for in the Act as Indigenous land use agreements (ILUAs) over specific areas of land and waters where native title has been determined to exist, where there are registered native title claimants, or where persons are claiming to hold native title. Under legal entitlement granted under the Native Title Amendment Act 1998, via these ILUA provisions (which will be discussed in detail on the following pages), resource development processes on Indigenous lands are usually negotiated directly between miners and native title claimants and holders for access to their lands.

ILUAs were designed as an alternative to the more formal legalistic native title process and were intended to provide for legally binding agreements between a native title claimant group and another party, thus creating certainty of tenure for pastoral and mining interests (Martin, 2015). The ILUAs scheme heralded a new age of agreement-making with Indigenous peoples, extending to native title claimants and holders the right to negotiate in good faith over future acts on their claimed lands, whether or not native title had been determined. ILUAs are often negotiated as part of the settlement of a native title determination application between the native title holders and the parties affected by the claim (such as pastoralists, energy and infrastructure providers, and local government).

Supporters of the ILUA scheme argue they are flexible and voluntary processes under which native title parties and companies can reach a legally binding agreement on a wide range of matters, including approval of future activities and multiple projects. Mining companies, for example, often prefer to use the ILUA process to provide for certainty of access for exploration and mining. Consequently, many agreements relating to mining projects in Australia are now processed as ILUAs, rather than under the "right to negotiate" (RTN) provisions of the NTA (Howlett and Lawrence, 2019). Thus, while there are those who contend that the amendments to the NTA in 1998 under the Howard Liberal government ushered in a more fruitful period of agreement-making, resulting in significant economic and social outcomes for native title parties (Langton and Webster, 2012, p. 79), others suggest that in reality, ILUAs reduced the bargaining position of native title claimants and holders and the potential for realizing new forms of governance on Indigenous lands (Howard-Wagner, 2010; Galloway, 2020; Howard-Wagner and Maguire, 2010).

In a qualitative study based on interviews with Indigenous stakeholders involved in ILUA negotiations, interviewees' documented concern about the absence of monitoring of outcomes from ILUA processes and questioned the potential of ILUAs to deliver any "practical" economic, social and cultural benefits (Howard-Wagner, 2010). The study concluded

> that there is presently little promise for land agreements, such as ILUAs, to be a panacea for the cycles of poverty and excessive welfare dependency among Indigenous peoples and local communities, such as creating sustainable regional "economic" development, or providing a mechanism for Indigenous peoples "to negotiate their way into the nation state, particularly in areas concerning land access, social and environmental management, and associated infrastructure development".
>
> (Howard-Wagner, 2010, p. 104)

This 2010 study preceded recent amendments to the NTA, the Native Title Amendment (Indigenous Land Use Agreements) Act 2017 (Cth), which embraces a weak approach to consent that further undermines collective rights in favor of representative rights, and thus further privileges Western decision-making processes over property settlement and law (Young, 2017).

The recent case of the ILUA agreement-making process for the Adani Carmichael Mine in Queensland is illustrative of the inherent problems within the Australian agreement-making process. The government approvals process for the mine has received intense media and environmental scrutiny as, once completed, the mine will be among the largest new coal mines in the world. The Galilee Basin is home to the Wangan and Jagalingou peoples (W&J), who in 2004 lodged a native title claim under the NTA. Although their claim has not yet been determined, the registration provided them the procedural rights in relation to activities that might affect their native title, such as the grant of mining and pastoral leases and other land uses. The W&J native title claim group rejected Adani's development proposals for the Carmichael Mine in December 2012, and again in October 2014 (Brigg, 2018). Despite their opposition to the project, in August 2014 the Queensland Coordinator-General recommended that the mining lease and Environmental Authority be granted. In April 2015, the National Native Title Tribunal (NNTT) found that the mining lease for the Carmichael Coal Mine could be granted by the Queensland government under Australia's NTA, despite the W&J withholding their agreement (Tauli-Corpuz, 2016). Several mining leases were issued to Adani by the Queensland government throughout the history of the development approval process—all of which are authorized by the NNTT and without the consent of all of the W&J native title claimants.

In 2016, a highly contested ILUA with Adani was signed by some of the W&J native title claimants, but significant concerns have been raised about its legitimacy. It was challenged in court by the W&J Family Council, a group of W&J traditional owners who have consistently opposed the development of the mine on their traditional lands. The W&J Family Council have continually sought to have this ILUA overturned, as the development does not have their consent and they were not signatories to the ILUA. In August 2018, the Australian Federal Court ruled in favor of Adani over the W&J Family Council, upholding the contested ILUA and paving the way for the Queensland government to cancel native title over the mine site (Robertson and Siganto, 2018). What this saga highlights is that the agreement-making process in Australia, under the current native title legislative framework, does not require the consent of all Indigenous parties whose native title rights will be affected by a development.

This failure to obtain the consent of *all* Indigenous native title claimants to legitimate an ILUA was recently addressed by the full bench of the Australian Federal Court in McGlade v Native Title Registrar (McGlade). The key issue addressed by the court was "whether an ILUA can be registered if not all individuals who jointly comprise the relevant registered native title claimant or claimants have signed the ILUA" (Frith, 2017, p. 25). The court held that certain types of ILUAs require *all* named registered native title claimants in a native title claim to execute the agreement. That decision cast doubt on the validity of previous ILUAs registered since 1998 in circumstances where not all of the registered claimants had signed the agreement (including the Adani ILUA). In response to this decision, the Australian Government quickly acted to amend the Native Title Act and passed the Native Title Amendment (Indigenous Land Use Agreements) Act 2017 in June 2017. Young (2017, pp. 26–34) argues that the amendments further infringe the rights of Indigenous peoples in Australia and may dispossess them as peoples, which is a community/collective rather than an individualistic/democratic notion, by using legal means to perpetuate a form of colonialism, and facilitating the construction of the pretense that they have in fact consented to their own dispossession.

Finally, in the current Australian agreement-making system under the NTA, there is no requirement that the claim group who sign an ILUA have a particular quorum present in order to ensure the "majority view" of the group present at an authorization meeting is in fact the majority view of the wider claim group. In addition, where a majority decision-making process is adopted according to Western standards of meeting procedure, as is most frequently the case, there is no provision for any cultural considerations whatsoever, such as the differences in authority between elders and young people (Forrest, 2017, p. 31). These inequitable characteristics of the architecture of the Australian agreement-making system privilege the interests of developers and subjugate the rights of Indigenous Australians to negotiate their free and informed consent to developments on their lands.

In summary, in Australia the institutional structures that characterize and support the native title system are still not conducive to ensuring that agreement-making delivers socioeconomic benefits to Indigenous Australians (Dillon, 2018). While there may be instances where Aboriginal agency can counterbalance the proclivities of these institutional structures, this agency is still circumscribed by a system that promotes neoliberal solutions to the questions of adjudication of property rights, and privileges the rights of corporate and landowning sectors. In the words of one of the original High Court Judges in the historic Mabo case, Justice McHugh, the system

> is stacked against the native title holders whose fragile rights must give way to the superior rights of the landholders whenever the two classes of rights conflict. And it is a system that is costly and time consuming. At present the chief beneficiaries of the system are the legal representatives of the parties. It may be that the time has come to think of abandoning the present system, a system that simply seeks to declare and enforce the legal rights of the parties irrespective of their merits.
>
> (McHugh, 2002, cited in Dillon, 2018, p. 9–10)

Agreement-making in Sweden

While agreement-making has had a much more substantial impact in Australia than in the Nordic states, it is gaining traction in Arctic regions, particularly in Sweden. Through an example of an agreement-making process in the wind power industry in Sweden, this section demonstrates that the Nordic states are not immune to mantras of marketization and privatization; in fact, if anything, they have in some ways been all the more willing to take on neoliberal logics and practices because of the critique they provide of the welfare state model, something all sides of politics have been keen to embrace in Sweden (Anttonen and Meagher, 2013). The case concerns an agreement not between a Sámi community and a resource developer (although these too are becoming all the more common in Sweden) but between a resource developer and the regional arm of the Swedish state.

The case is about a proposed wind power development in Stekenjokk in the North of Sweden: formally on Crown Lands on the Swedish side of the

Scandinavian mountain chain and just a few kilometers from the Norwegian border. It is an area customarily used by Sámi people for reindeer husbandry since time immemorial. During 2008, the regional state authority in mid-northern Sweden—the Västerbotten County Administrative Board—invited three private wind power companies to tender for the exclusive right to explore the feasibility of a wind power park. A Letter of Intent was later signed between the successful bidder, Fred Olsen Renewables (FOR), a Norwegian wind power company, and the County Board. The tendering process and the Letter of Intent were framed by the County Board as a routine administrative solution to the increased interest and pressure from multiple wind power companies wanting to develop the same area. Yet, there was nothing routine about it. In the Letter of Intent, the County Board also outlined its intention to negotiate a profit-based land lease agreement with FOR, if the project were given planning permission. This is where the problems began: a County Board had never previously attempted such a "market solution" and, as the extended arm of the Swedish state, it caused considerable controversy.

This was also in the context of increasing environmental and public opposition to wind power in Sweden since the early 2000s, whereby several local municipalities had begun to exercise their veto power to block projects. Moreover, the institutional and legal context is complex and uncertain, with extensive opportunities for appeal: Sweden has a relatively long average lead time of over five years (Pettersson et al., 2015, p. 3122) for new wind power projects. Direct negotiations between proponents and affected parties have thus emerged as a central solution so as to avoid prolonged appeals. This has increasingly taken the shape of wind power proponents negotiating benefit-sharing and part-ownership agreements with local municipalities, landowners and neighboring (non-reindeer herding) local communities.

In this context, the status of Sámi reindeer herding communities in such negotiations has been contested as they have often been completely marginalized from the process. While wind power developers have recognized landowners as a necessary negotiating party—both private landowners and the state as "owner" of Crown lands—the same recognition has not been historically extended to Sámi communities. Reindeer grazing rights are a user right that burden vast areas in the north, on both state and privately-owned lands. However, Sámi land uses have been commonly rendered invisible, thus also making legitimate the argument that there is, quite simply, more room for wind power "up north" than in the more heavily populated and industrialized southern areas of Sweden. Also at work is the pervasive idea that Sámi traditional land uses are infinitively "adaptable" and can peacefully exist in parallel with other competing land uses, which again simply ignores the harms suffered by reindeer herders as they are forced to buffer against an exponential number of intrusions onto their traditional lands (Lawrence and Åhrén, 2016). As wind power developments have boomed, two interconnected forces have thus been at work. One is the increased need for wind power developers to access land; the other is the increased resistance by Sámi communities to the paternalistic role of the state and the systemic marginalization of Sámi rights and interests in planning processes.

Sámi communities and organizations have therefore progressively sought to bypass state permitting processes and exercise a Sámi right to self-determination by engaging proponents in direct negotiations. Increasingly, these have resulted in benefit-sharing agreements between wind power companies and Sámi communities. These generally include both compensation for direct damages (which may include lost pasture, but also compensation for extra feeding, animal transportations, etc.) as well as a set percentage of production per windmill (with a minimum rate guaranteed). In addition, some communities have also negotiated a direct payment per windmill upon construction. In some cases, these agreements have provided Sámi communities with unprecedented economic opportunities and allowed them to invest, among other things, in collective infrastructure for their reindeer herding activities. However, given the fact that these negotiations take place in a legal and political context in which Sámi rights are systematically ignored and sidelined in land use planning processes (Raitio, Allard and Lawrence, 2020), many of these same communities also concede that this places them in a catch-22 of two kinds. First, they are forced to negotiate agreements to projects they essentially opposed, because if they do not, they risk the project going ahead anyway, but without any decent compensation. Second, they become ever more dependent upon yet more compensations and profits from resource development in order to pay for the mitigation measures necessary in order to make reindeer husbandry viable (e.g., more fencing, more active herding measures, more artificial feeding, etc.).

In the absence of an Indigenous veto over developments, negotiations are rarely "freely" entered into by either party. Wind power companies, in most cases, are only willing to come to the negotiating table *after* they are exposed to external pressure. Local Sámi communities and Sámi organizations—such as the National Association of Swedish Sámi (Svenska Samernas Riksförbund, SSR) and the Saami Council—have used various "arts of resistance" (Tully, 2000) to exert such pressure, including local Sámi protests over developments; investor activism (Lawrence, 2008); the filing of complaints to the UN Committee on the Elimination of Racial Discrimination and to member states under the OECD Guidelines for Multinational Corporations; and campaign partnerships with international NGOs, such as Greenpeace (Lawrence, 2008) and BankTrack. Moreover, for many Sámi communities, the ultimate goal is not dialogue, but to protect remaining reindeer grazing lands from further industrial encroachments, given the already enormous pressure on traditional Sámi lands from forestry, hydropower, mining, roads, tourism and other infrastructure. In the face of a general lack of respect for Sámi user rights, a negotiated outcome is sometimes the most realistic goal for affected communities. There is only one known example of a genuinely equal partnership where both a Sámi community and wind power company have freely entered into a partnership to plan for a wind power project on Sámi territories. Ironically, this project ultimately failed to gain planning approval because the planning authorities deemed the area to have high environmental values, but for Sámi communities, they viewed the area of little value because of existing hydro developments.

In the case of the Stekenjokk agreement between the County Board and Fred Olsen Renewables, the local Sámi community of Vilhelmina Norra was not party to the agreement nor to any of the negotiations. Instead, the Swedish state played both the role of paternal protector by claiming to represent Sámi interests in the negotiations, but also simultaneously the role of market actor by staking a claim to a market share of any profits. These maneuverings suggest that the state's concern is simply to create a fair market of competition between wind power proponents and landowners. They also link into a broader neoliberal discourse that presumes that the market can, and should, provide solutions to all manner of problems (Dean, 1998).

Market rationalities reconfigure the Indigenous–state relationship in specific ways. They simultaneously reproduce inequalities and depoliticize the power relations producing those inequalities. Here, governing activities are "recast as non-political and non-ideological problems that need technical solutions" (Ong, 2006, p. 3), and the inherently political nature of resource extraction in Indigenous territories is left aside. The Letter of Intent was drawn up by the state, without any Sámi participation, yet the politics of this exclusion is rendered invisible through the claim that the state, performing as a market actor in "market negotiations," is not required to consult with affected communities. This echoes the very same logic underpinning the dispossession of the Sámi of their taxed lands during the nineteenth century: fundamental shifts in Sámi land rights were framed by the state as mundane "reindeer herding administrative issues." Bureaucratic and "everyday" as they are, "it is these putatively technical and unremarkable practices that render tenable the political tasks of state formation, governance, and the exertion of power" (Sharma and Gupta, 2006, p. 11). In short, mundane bureaucratic practices of the state function to make the technical government of Indigenous peoples "apolitical" (Sharma and Gupta, 2006, p. 11): they have "anti-political" effects because they seek to limit contestation, debate and protest (Barry, 2002).

Moreover, at the same time as the familiar discourse of protectionism is mobilized to marginalize Sámi people from decisions affecting them, the practice of *actual* protection of Sámi areas is being increasingly undermined as the state seeks to actively promote development above the cultivation border. In this context, an environmental discourse concerning sustainable energy is used to justify the state's facilitation of wind power developments in the mountains. Sámi land uses and occupation of these same lands are thus viewed as a hindrance to the state's renewable energy goals, and negotiated agreements can be viewed as facilitating yet another form of accumulation by dispossession (Howlett and Lawrence, 2019).

Conclusion

The preceding discussion has attempted to unpack the idea of agreement-making as a positive beneficial development in land and marine use governance—a trope that posits agreements as invariably beneficial for, and demonstrative of, Indigenous agency. There is no doubt that Indigenous communities and organizations exert their agency in a myriad ways, and for particular communities, negotiated agreements have provided opportunities and benefits to which they may not previously have had access. Those benefits acquired under native title arrangements are not because corporations spontaneously decided to give graciously and generously, but because Indigenous communities and organizations have had to fight hard and long for just agreements by strategically engaging with the corporate, legal and political spheres in complex and savvy ways. The fact remains that the adverse impacts of resource developments on Indigenous lands overwhelmingly tend to outweigh the benefits, even when the best negotiated agreements are in place. A nuanced critique of negotiated agreements and the efficacy of agreement-making as a land use governance tool in terms of outcomes for Indigenous peoples is not the same as denying that Indigenous agency exists.We argue that agency without autonomy and sovereignty will always be circumscribed by the structural, institutional and colonial/historical realities within which that agency is expressed.Via the abrogation of Indigenous rights through the mirage of "agreement-making," Indigenous peoples are forced into an extremely limited form of "agency," one where they are forced to engage with, and consent to, tools that ultimately dispossess them (Howlett and Lawrence, 2019).

The neoliberal logic that informs agreement-making thus takes contentious resource developments and makes them "more easily 'governable' for both governments and companies by supposedly shifting from adversarial politics to consensual agreements" (Peterson St-Laurent and Billon, 2015, p. 592). At the risk of repetition, we argue that agreements confer consent, consent then is presented as evidence of agency and proof of Indigenous capacity in these agreements. By some sleight of hand, agreement-making is thus construed as addressing Indigenous rights, seemingly because Indigenous peoples entered into an agreement about those rights.Yet agreement-making has not manifested out of a recognition of Indigenous rights, as detailed in the previous pages, but rather as a means to abrogate those rights under the guise of manufactured consent.

The discussion presented in this chapter does not seek to undermine, negate or romanticize Indigenous agency as it manifests in this neoliberal milieu of agreement-making. On the contrary, it suggests that the instances where Aboriginal peoples have managed to secure positive economic and social outcomes from agreement-making have been largely due to their own agency (see O'Faircheallaigh, 2016). However, this agency is always circumscribed by the structural realities of the institutional and political/economic architecture that underpins state claims to sovereignty, and that has established agreement-making as the norm in resource governance regimes.

Dismissing the reality that neoliberalism structures the agreement-making process in resource governance processes—which sets aside the question of consent and leaves Indigenous peoples with little choice but to negotiate—is unjust and can further serve to dispossess Indigenous peoples of both their rights and their lands. These tools are developed to minimize disruption to business as usual, and benefit corporations and ultimately the state. The state abrogates the business of protecting Indigenous rights to the agreement-making process, and Indigenous peoples who participate in these processes (often with no alternative), are seen to be demonstrating their agency. Part of the discursive and ideological power of neoliberalism is that many of the key processes and policies that have emanated from it are normalized to the extent that to question them is deemed irrational, ideological or extreme. At the same time, neoliberalism itself tends to become invisible and the policies that emerge from it, such as agreement-making, are suddenly seen as natural and common sense.

Finally, the key insight that is critical for those Sámi communities who contemplate entering into agreement-making regimes in Norway is that *agreement-making is not a panacea for just, inclusive treatment of Indigenous rights in land and marine use governance processes.* On the contrary, agreement-making is a neoliberal governance tool that obfuscates the decision-making processes for land, marine and resource development processes and presents Indigenous participation in these processes as evidence of their consent. Indigenous agency in this milieu is then deemed the measure of the success of agreement-making. This is both disturbing and disingenuous. The danger in an uncritical acceptance of agreement-making as the panacea for Indigenous peoples is that it risks promoting neoliberal practices, which ultimately serve to further dispossess them. In those cases where Indigenous rights are significantly impacted, and in the absence of an Indigenous veto over resource developments, there is no such thing as a fair and just negotiated agreement. Only small and constrained maneuverings can exist, whereby Indigenous peoples try to get the best out of a bad deal.

References

Anttonen, A. and Meagher, G. (2013) "Mapping marketisation: concepts and goals," in G. Meagher and M. Szebehely (eds.) *Marketisation in Nordic eldercare*. Research Report. Stockholm: Department of Social Work, Stockholm University, pp. 13–22.

Bargh, M. (2007) "Introduction," in M. Bargh (ed.), *Resistance: an Indigenous response to neoliberalism*. Wellington: Huia Books, pp. 1–21.

Barry, A. (2002) "The anti-political economy," *Economy and Society*, 31, pp. 268–284.

Brigg, M. (2018) "Native title colonialism, racism and mining for manufactured consent," *New Matilda*, January 30 [online]. Available at: https://newmatilda.com/2018/01/30/native-title-colonialism-racism-adani-and-the-manufacture-of-consent-for-mining/ (Accessed April 29, 2018).

Brown, W. (2015) *Undoing the demos: neoliberalism's stealth revolution*. Brooklyn, NY: Zone Books.

Cahill, D. (2007) "The contours of neoliberal hegemony in Australia," *Rethinking Marxism*, 19(2), pp. 221–233.

Cameron, E. and Levitan, T. (2014) "Impact and benefit agreements and the neoliberalization of resource governance and Indigenous–state relations in northern Canada," *Studies in Political Economy*, 93(1), pp. 25–52.

Dean, M. (1998) "Administering asceticism: reworking the ethical life of the unemployed citizen," in M. Dean and B. Hindess (eds.) *Governing Australia: studies in contemporary rationalities of government*. Cambridge: Cambridge University Press, pp. 87–107.

Denniss, R. (2018) "Dead right, how neoliberalism ate itself and what comes next," *Quarterly Essay*, 70.

Dillon, M. (2018) "Systemic innovation in native title," *Discussion Paper No. 294*. Centre for Aboriginal Economic Policy Research, School of Social Sciences College of Arts & Social Sciences. Canberra: Australian National University.

England, K. and Ward, K. (2007) *Neoliberalization: states, networks, peoples*. Oxford: Blackwell.

Forrest, K. (2017) "A Wajuk Barladong Mineng Nyungar perspective on McGlade v Native Title Registrar and the resulting Native Title Amendment (Indigenous Land Use Agreements) Bill 2017," *Indigenous Law Bulletin*, 8(28), pp. 29–31.

Frith, A. (2017) "Case note: McGlade v Native Title Registrar," *Indigenous Law Bulletin*, 8(28), pp. 24–28 [online]. Available at: http://www5.austlii.edu.au/au/journals/IndigLawB/2017/8.html (Accessed May 20, 2020)

Galloway, K. (2020) "Courts of the conqueror: Adani and the shortcomings of native title law," *AQ: Australian Quarterly*, 91(1), pp. 14–20 [online]. Available at: https://research-repository.griffith.edu.au/bitstream/handle/10072/391049/Galloway348490-published.pdf?sequence=3&isAllowed=y (Accessed November 19, 2020)

Harvey, D. (2010) *The enigma of capital and the crisis of capitalism*. London: Profile Books.

Heynen, N., McCarthy, J. and Prudham, W. (2007) *Neoliberal environments: false promises and unnatural consequences*. London and New York: Routledge.

Howard-Wagner, D., Bargh, M. and Altamirano-Jiménez, I. (2018) "From new paternalism to new imaginings of possibilities in Australia, Canada and Aotearoa/New Zealand: Indigenous rights and recognition and the state in the neoliberal age," in D. Howard-Wagner, M. Bargh, and I. Altamirano-Jiménez (eds.) *Neoliberal state, recognition and Indigenous rights: new paternalism to new imaginings*. Canberra: Australian National University, pp. 1–40.

Howard-Wagner, D. (2010) "Scrutinising ILUAs in the context of agreement making as a panacea for poverty and welfare dependency in Indigenous communities," *Australian Indigenous Law Review*, 14(2), pp. 100–114.

Howard-Wagner, D. and Maguire, A. (2010) "'The holy grail' or 'the good, the bad and the ugly'? A qualitative exploration of the ILUAs agreement-making process and the relationship between ILUAs and native title," *Australian Indigenous Law Review*, 14(1), pp. 71–86.

Howlett, C., Seini, M., McCallum, D. and Osborne, N. (2011) "Neoliberalism, mineral development and Indigenous people: a framework for analysis," *Australian Geographer*, 42(3), pp. 309–323. doi:10.1080/00049182.2011.595890

Howlett, C. and Lawrence, R. (2019) "Accumulating minerals and dispossessing Indigenous Australians: native title recognition as settler-colonialism," *Antipode*, 51(3), pp. 818–837.

Langton, M. and Webster, A. (2012) "The 'right to negotiate,' the resources industry, agreements and the Native Title Act," in T. Bauman, and I. Glick (eds.) *The limits of change: Mabo and native title 20 years on*. Canberra: Australian Institute of Aboriginal and Torres Strait Islander Studies (AIATSIS) Research Publications.

Larsen, D.R. (2018) "Reindeer owners say no to 122 million kroner," *NRK*, September 6.

Lawrence, R. (2008) "NGO campaigns and banks: constituting risk and uncertainty," *Research in Economic Anthropology*, 28, pp. 241–269.

Lawrence, R. and Åhrén, M. (2016) "Mining as colonisation: the need for restorative justice and restitution of traditional Sami lands," in I. Head, K. Saltzman, G. Setten, and M. Stenseke (eds.) *Nature, temporality and environmental management: Scandinavian and Australian perspectives on peoples and landscapes*. London: Routledge, pp. 149–166.

McCormack, F. (2017) *Private oceans: the enclosure and marketisation of the seas*. London: Pluto Press.

Maddison, S. (2013) "Indigenous identity, 'authenticity' and the structural violence of settler colonialism," *Identities*, 20(3), pp. 288–303. doi:10.1080/1070289X.2013.806267

Martin, D.F. (2015) "Does native title merely provide an entitlement to be native? Indigenes, identities, and applied anthropological practice," *Australian Journal of Anthropology*, 26(1), pp. 112–127.

O'Faircheallaigh, C. (2016) *Negotiations in the Indigenous world: Aboriginal peoples and the extractive industry in Australia and Canada*. New York: Routledge.

Ong, A. (2006) *Neoliberalism as exception: mutations in citizenship and sovereignty*. Durham, NC: Duke University Press.

Papillon, M. and Rodon, T. (2019) "The transformative potential of Indigenous-driven approaches to implementing free, prior and informed consent: lessons from two Canadian cases," *International Journal on Minority and Group Rights*, 27, pp. 314–335.

Papillon, M., Leydet, D. and Leclair, J. (2020) "Free, prior and informed consent: between legal ambiguity and political agency," *International Journal on Minority and Group Rights*, 27(2), pp. 1–10.

Peterson St-Laurent, G. and Billon, P.L. (2015) "Staking claims and shaking hands: impact and benefit agreements as a technology of government in the mining sector," *Extractive Industries and Society*, 2(3), pp. 590–602 [online]. doi: 10.1016/j.exis.2015.06.001

Pettersson, M., Ek, K., Söderholm, K. and Söderholm, P. (2015) "Wind power planning and permitting: comparative perspectives from the Nordic countries," *Renewable and Sustainable Energy Reviews*, 14, pp. 3116–3123.

Pinkerton, E. and Davis, R. (2015) "Neoliberalism and the politics of enclosure in North American small-scale fisheries," *Marine Policy*, 61, pp 303–312.

Polanyi, K. (1957) *The great transformation. The political and economic origins of our time.* 2nd ed. Boston: Beacon Press.

Raitio, K., Allard, C. and Lawrence, R. (2020) "Mineral extraction in Swedish Sápmi: the regulatory gap between Sami rights and Sweden's mining permitting practices," *Land Use Policy*, 99, article 105001.

Rigby, B., Davis, R., Bavington, B. and Baird, C. (2017) "Industrial aquaculture and the politics of resignation," *Marine Policy*, 80, pp. 19–27.

Robertson, J. and Siganto, T. (2018) "Adani Indigenous challenge dismissed by Federal Court, Government could cancel mine native title," *ABC News Online*, August 17. Available at: http://www.abc.net.au/news/2018-08-17/adani-federal-court-traditional-owners-native-title/10131920 (Accessed October 12, 2018)

Sharma, A. and Gupta, A. (2006) "Introduction: rethinking theories of the state in an age of globalization," in A. Sharma and A. Gupta (eds.) *The anthropology of the state: a reader.* Malden, MA; Oxford; Carlton, Victoria: Blackwell Publishing, pp. 1–42.

Stanner, W. (1979) *White man got no dreaming: essays 1938–1973.* Canberra: Australian National University Press.

Strakosch, E. and Macoun, A. (2012) "The vanishing endpoint of settler colonialism," *Arena Journal*, 37/38, pp. 40–62.

Sullivan, P. (2018) "The tyranny of neoliberal public management and the challenge for Aboriginal community organisations," in D. Howard-Wagner, M. Bargh, and I. Altamirano-Jiménez (eds.) *The neoliberal state, recognition and Indigenous rights: new paternalism to new imaginings.* Canberra: ANU Press, pp. 201–215.

Tauli-Corpuz, V.L. (2016) *Mandate of the special rapporteur on the rights of Indigenous peoples*, February 29 [online]. Available at: https://spcommreports.ohchr.org/TMResultsBase/DownLoadPublicCommunicationFile?gId=13727 (Accessed February 13, 2018)

Tully, J. (2000) "The struggles of Indigenous peoples for and of freedom," in D. Ivison, P. Patton and W. Sanders (eds.) *Political theory and the rights of Indigenous peoples.* Cambridge: Cambridge University Press, pp. 26–59.

Wishart, D.A. (2005) "Contract, oppression and agreements with Indigenous people," *University of New South Wales Law Journal*, 46(28), pp. 780–799.

Young, S. (2017) "Native Title Amendment (Indigenous Land Use Agreements) Act 2017 (Cth): relying on human rights to justify a legalised form of colonial dispossession?" *Indigenous Law Bulletin*, 8(30), pp. 24–28.

9 Paradigm conflicts

Challenges to implementing Indigenous rights in Sápmi

Kaja Nan Gjelde-Bennett

Introduction

Over thirty years in the making, the final draft of the Nordic Sámi Convention (NSC) was released in January of 2017. The NSC initially aimed to give the same basic rights to Sámi in Norway, Sweden and Finland, involving the Nordic state governments, the Sámi Parliaments of the three countries, the Saami Council and the intergovernmental Nordic Council. However, the NSC's final draft has been criticized for not realizing what it initially proposed to do, to unite the Sámi across state borders and secure their collective rights, including rights to land and natural resources under international law.

Meanwhile, debates on Indigenous rights continue to erupt into various conflicts across the Nordic states, involving industry (oftentimes international corporations), state governments and Sámi communities. The Kallak mine controversy in northern Sweden exemplifies one such confrontation. Tensions have been building for years between the Beowulf Mining company, the municipal and national governments and the Sámi communities in Gállok, in Jokkmokk Municipality, Sweden. Local government officials favor the proposed mine, citing the international mining company's promise of economic growth. Members of the Sámi communities and their allies remain vehemently opposed to the adverse environmental impacts of the industrial project, citing national and international law that grants the Sámi access to reindeer grazing land for the protection of their culture and traditional livelihoods.

Indigenous scholar Jerry Mander (2006, p. 4) identifies these phenomena as "paradigm wars," or conflicts involving different views of reality, usually concerning Indigenous rights to territory and natural resources. The NSC's failure to bolster Indigenous rights in the Nordic countries highlights the proliferation of industrial projects in the Arctic, which Sámi rights activists hoped to address through the transnational agreement (Gjelde-Bennett, 2017). Utilizing the concept of paradigm conflicts as a primary theoretical lens, this chapter will explore Indigenous rights debates in global politics theoretically and empirically to expose the structural inequalities within the international system that hinder the advancement of Indigenous rights in the Nordic area.

After defining the neoliberal and Indigenous paradigms, the case study of the proposed Kallak mine in Gállok, Sweden demonstrates the reality of paradigm

DOI: 10.4324/9781003131274-9

conflicts within the Nordic countries, as the debate centers around diverse ways of viewing the land in Gállok, how it should be used, and who has the right to make those decisions. Following the TriArc model of Arctic governance (Hernes, Broderstad and Tennberg, this volume), the interactions between local and state governments, industrial companies and Sámi communities in Gállok clearly illustrate the unequal power structures that deny the Sámi basic Indigenous rights granted by national and international law. These inequalities and injustices catalyzed the creation of the Nordic Sámi Convention as a channel to resolve ongoing conflicts particularly concerning natural resource management in Sápmi. Though the NSC in its current form has failed to settle ongoing paradigm conflicts such as that in Gállok, it remains compelling evidence of changing norms in the international system for addressing these structural inequalities to realize essential Indigenous rights to self-determination and natural resources.

Paradigm conflicts

To understand why the NSC failed to realize its ambitions to effectively address conflicts over natural resources in the Arctic, one must examine not only the regional context of the Nordic countries, but also the overarching power structures within global politics which necessitated the formation of the NSC. Namely, the prevailing international system dominated by neoliberal values upholds the ultimate authority of the sovereign state despite the proliferation of non-state actors. Representing a minority in most states within a neoliberal-dominated international system, Indigenous actors assuming an Indigenous worldview are often at a disadvantage for representing their own interests in global politics. As a result, there exists an imbalance of power between state governments, industries and Indigenous actors, characterized by the neoliberal and Indigenous paradigms within the international system. This inequality influences the structure and practice of global politics to prioritize state sovereignty over Indigenous rights (Gjelde-Bennett, 2017).

The neoliberal and Indigenous paradigms will first be defined from the disciplines of international relations and Indigenous studies respectively to establish the two main differing approaches to global politics. Theoretically framing Indigenous rights debates as opposing paradigms exposes historical structural inequalities within the international system which favor the neoliberal paradigm, while the incorporation of the Indigenous paradigm acknowledges the possibility of reconciling the two diverse worldviews.

The neoliberal paradigm

During the Cold War, the International Relations theory of neoliberalism was developed in response to increased globalization and the emergence of a plethora of nongovernmental (NGOs) and intergovernmental (IGOs) organizations since the Second World War. Neoliberalism draws its conceptual base from classical liberalism, which champions democracy, free trade and universal human rights. Neoliberalism follows the liberal basic theoretical assumption that the international system is anarchic and sovereign states are the most powerful and essential international actors (Keohane, 1984). Furthermore, neoliberalism conceptually explains the role of international regimes and institutions, which are formed to facilitate multilateralism for managing the anarchic international system to maximize absolute gains. International institutions incorporate norms, rules and practices that dictate behavioral roles, constrain activity and influence expectations, while regimes represent decision-making procedures around which actors' expectations converge in a particular area of global politics (Gjelde-Bennett, 2017). Usually comprised of legally non-binding agreements, international regimes provide rules of thumb, which create greater security in the anarchic international system by making state behavior more predictable and by compensating for individual limited rationality. This encourages cooperation among states without challenging state sovereignty (Keohane, 1984).

Post-WWII, the liberalism of privilege perspective spurred global powers such as the United States and other "Western" states to spread liberal values of democracy, multilateralism and free trade internationally. The hegemony of these Western democracies made other, less powerful countries quick to adopt liberal values, and then later neoliberal institutional relationships to participate in the international system. Within the more contemporary context of globalization, neoliberal institutions and regimes have been effectively disseminated throughout the international system by powerful actors, including states, IGOs, NGOs and transnational corporations. Thus, neoliberalism has become the dominant paradigm of global politics today.

In their work *Neoliberalism and post-welfare Nordic states in transition*, Guy Baeten, Anders Lund Hansen and Lawrence D. Berg (2016) argue that—specific to contemporary Nordic states—these countries are impacted continuously by evolving processes of "neoliberalization." They submit that the Nordic states are in a "post-welfare" phase; while the state governments maintain welfare state policies, they are experiencing a shift in their priorities away from the welfare state model in favor of organizing state provisions "more in line with market principles" (Baeten, Lund Hansen and Berg, 2016). The proliferation of industrial projects in the Arctic corroborates the continued prevalence of core neoliberal norms and values in the Nordic countries in conjunction with welfare state policies. Neoliberalism has not disappeared, merely taken on new forms.

The Indigenous paradigm

The Indigenous paradigm does not claim to represent the perspectives of all Indigenous peoples, but for the purpose of this chapter, it is principally informed by one of the main subdivisions of thought within Indigenous studies: indigenism. As Indigenous peoples have historically been excluded from creating research agendas in academia for studying Indigenous peoples, they have also been marginalized in political processes concerning their own communities that threaten their collective rights. Therefore, Indigenous scholars assuming an indigenist perspective argue for an Indigenous paradigm allowing Indigenous peoples to construct their own systems of gaining and synthesizing knowledge. In a collaborative anthology on emerging theoretical lenses within Indigenous research, Gerald Roche, Åse Virdi Kroik and Hiroshi Maruyama (2018, p. 229) provide a more holistic definition:

Indigenism refers to the transnational movement to promote the political interests of Indigenous people, including promotion of the universal applicability of the categories "Indigenous peoples" and "indigeneity" (Niezen 2003; Clifford 2013). Although originating primarily in the CANZUS (Canada, Australia, New Zealand and the US) countries, and now vigorously supported by the Nordic countries (Finland, Sweden and Norway), this movement has since taken on global dimensions (Merlan 2009), including the creation of legal norms and international agreements that have developed constant feedback between local movements and global networks.

To expand upon the principal characteristics of indigenism, Cree scholar Margaret Kovach's work on developing Indigenous research methodologies explains an Indigenous approach as holistic and based on the relational knowledge production that aims for reciprocity rather than extracting resources (Kovach, 2010). It follows that the Indigenous paradigm's core values are founded on relational accountability. The most important aspect of this is role fulfillment for maintaining productive relationships that include respect, reciprocity and responsibility (Gjelde-Bennett, 2017). In academia, Indigenous research approaches are designed to shift the power from the researcher to the participants, ensuring the project gives back to the Indigenous community. The decentralization of power acknowledges participants' agency within the research process and includes individuals' subjective experiences (Gjelde-Bennett, 2020). Indigenous research is also context-specific and centered around a particular tribal epistemology (Kovach, 2010). For instance, Cree knowledges and ways of knowing are distinct from those of the Sámi because they are based within a particular ontology (Kovach, 2010). Kovach (2010, p. 61) clarifies that "Indigenous epistemologies assume a holistic approach that finds expression within the personal manifestations of culture." Therefore, an indigenist approach, within or outside of an academic context, centralizes Indigenous ontologies and epistemologies which vary depending on the Indigenous context being examined.

Moreover, the Indigenous paradigm is made up of systems of knowledge formed by the relationships between various subjects (people, objects, concepts) and therefore represents a source of knowledge as well as a way of knowing (Gjelde-Bennett, 2017). According to Indigenous scholar Shawn Wilson (2008, p. 74), "Indigenous epistemology is our system of knowledge in their context, or in relationship," which includes Indigenous cultures, histories and worldviews. Wilson (2008) contends that multiple possible realities exist which differ depending on one's perspective, and that knowledge itself is not the final goal but the potential for change that knowledge holds. In other words, "reality is not an object but a process of relationships" (Wilson, 2008, p. 73). Similarly, Indigenous rights activists La Donna Harris and Jacqueline Wasilewski (2004, p. 498) contend that in "the Indigenous perspective it is assumed that we have each had different experiences, so… there are multiple realities" (Harris and Wasilewski, 2004, p. 498). Additionally, relational knowledge does not just originate from interpersonal relationships, but is shared with the natural world to include "all of creation," which transcends the idea of "individual knowledge to the concept of relational knowledge" (Wilson, 2008, p. 74).

Within the context of global politics, the Indigenous paradigm invites an understanding of how the international system is being created by relationships between actors, conceptual frameworks and the environment. Furthermore, Canadian Indigenous scholar and activist Taiaiake Alfred argues that the Indigenous approach to global politics features a regime of respect, which stands in contrast to the presumed superiority of the state (Lightfoot, 2016). In Sheryl Lightfoot's work on Indigenous politics, Alfred (in Lightfoot, 2016, p. 10) asserts, the imperative of respect precludes the need for homogenization, and thus realizes the potential for more peaceful relations. As opposed to the top-down process of decision-making in the neoliberal paradigm with powerful states as the most important actors, the Indigenous paradigm argues for a more consensus-based, bottom-up approach that incorporates multiple perspectives.

Specific to conflicts over natural resources, Winona LaDuke of the White Earth Land Recovery Project emphasizes the importance of place to Indigenous ontologies. Indigenous peoples do not own land in terms of legal property rights as defined by a Western neoliberal lens. Instead, Indigenous peoples belong to the land and have a relationship with the land, which innately includes certain roles and responsibilities to respect and protect it (Mander, 2006). This relational ontology within an Indigenous paradigm leads to a distinct understanding of development and sustainability. In their article on creating more inclusive definitions of environmental sustainability, Indigenous studies scholars Pirjo Kristiina Virtanen, Laura Siragusa and Hanna Guttorm (2020, p. 80) contend that the UN's Sustainable Development Goals should not be universally implemented due to the fundamental ontological and epistemological differences between peoples. Focusing on Indigenous perspectives, they state:

> Indigenous approaches to sustainability and development emphasize place and locality, relationships and sacred exchanges, where the quality of life is measured and adjusted to meet the needs of the human non-human community and future generations.

The Kallak mining project controversy

The heated controversy over the Kallak mining project in Sweden exemplifies how individual Indigenous rights conflicts over natural resource management are inextricably linked to larger paradigm conflicts within global politics. Informed by the TriArc model of Arctic governance, this section will detail the evolving legal debate between governmental, industrial and Indigenous actors concerning the construction of an iron ore mine in Gállok. Then the case study will be examined from the neoliberal and Indigenous paradigms to expose the contentious conceptual underpinnings of this conflict that produce inequalities within the international system.

Case study

In 2010, the United Kingdom-based Beowulf Mining company and its Swedish subsidiaries began applying for exploration permits for a mining project in Gállok.

Gállok (Kallak in Swedish) is located in Jokkmokk Municipality, a community with Indigenous and non-Indigenous residents in Sweden's northernmost Norrbotten County. Sixty-three percent of reindeer herders in Sweden reside in this county, with 32 groups of reindeer herding enterprises (Pantzare Information AB, 2010). The proposed mine is within close proximity to the Sirges and Jåhkågasska Sámi herding communities, who are concerned the mining project will endanger their livelihood (Solly, 2016).

Beowulf Mining Plc advertises the proposed mine as a unique opportunity to stimulate the local economy, estimating over one hundred million metric tons of ore could be mined in Kallak (Beowulf Mining, 2020). In Truls Anderssen's documentary film *Gállok* (2019), Kjell Ek provides his more localized perspective on the proposed mining project as a miner in northern Sweden:

> The indirect consequence [of the mine] would be that the entire community grows… I think about 100 people disappear from Jokkmokk every year… In ten years' time, we'll be fewer than 4,000 inhabitants… All these public utilities are dependent on there being people around.
>
> (*Gállok*, 2019)

Generally, local non-Indigenous residents support the mine for facilitating economic growth, while the Sámi communities claim the mine would both geographically divide the Jåhkågasska Sámi community and disrupt overall reindeer herding activities (*Gállok*, 2019). The proposed mining site in Gállok rests on herding land used for winter grazing, and its full construction would reduce the reindeer's territory and natural food supply (Forss, 2015). Limiting access to winter grazing land violates the national Reindeer Husbandry Act of 1971 that gives Sámi reindeer herders access to traditional lands and the natural resources within them, regardless of land ownership, for the purpose of herding and to preserve their culture (Williams, 2003). On the international level, both the International Labour Organization Convention No. 169 (ILO 169) and the United Nations Conference on Environment and Development (UNCED) Agenda 2151 grant Indigenous peoples control over their traditional territories and the natural resources within them for protecting their livelihoods and promoting sustainable development. Despite international pressures, Sweden has yet to ratify ILO 169, and the UNCED recommendations are legally non-binding (Anaya, 2011). Additionally, the local Sámi communities and other critics of the mining project are concerned about the environmental impact of the mine despite Beowulf's assurances the mine will be sustainably developed.

To combat the construction of what they believe to be an unlawful mine, the Sámi communities around Jokkmokk have led a series of peaceful protests outside the mining site since 2011, some of which resulted in the forced removal of demonstrators by Swedish police in July and August of 2013 (Solly, 2016). According to the Sámi communities, Beowulf falsely claimed to have included the Sámi "villages" in the environmental impact assessment during the Raw Materials Group (RMG) Mining Investment Conference in November of 2011 (Solly, 2011).

During a video-recorded stakeholders' meeting in Stockholm, former CEO of Beowulf Mining Plc, Clive Sinclair-Poulton, went so far as to claim that there were no people living near the proposed mine:

> One of the major questions I get is what are the local people going to go ahead and say about this project, I show them this picture [landscape photograph with mountains and forest] and I say what local people?
>
> (*Gállok*, 2019)

In December 2011, Beowulf was caught drilling with an expired permit and was reported again in January of 2012 for drilling illegally in breach of the Swedish Mining Act. Citing international law and the Swedish Constitution in response to the mining company's behavior, the Sirges and Jåhkågasska Sámi communities stated in an open letter to Clive Sinclair-Poulton that the company's disrespect for Sámi wishes in continuing to pursue the mining project in Gállok was a violation of their rights to the land and their human rights to culture and to health, and that they did not intend to cooperate with the company (Solly, 2012).

Meanwhile, Beowulf continues to prepare the Kallak mining project and awaits the Swedish government's approval for its 25-year concession (Solly, 2012). Though the Swedish government has not been openly opposed to the project and the Jokkmokk mining inspectorate has approved the mine, the Swedish government has expanded the criteria for the environmental impact assessment, obliging Beowulf to meet further environmental requirements before its concession can be approved (Boland, 2016).

Despite the multiple international and national law bodies cited by the Sámi communities and their allies, the Sámi in Jokkmokk not only lack legal rights to the land, but also lack political representation in the local and state governments to influence the Swedish government's decision-making (Gjelde-Bennett, 2017). There is still a distinct chance the Kallak mine will be constructed. This case reveals the inadequacies of Swedish government policy and the limited powers of the Sámi communities to protect their fundamental rights. For the Sámi, the construction of the mine would not just hurt the reindeer herding industry, it would also undermine their human and Indigenous rights. It appears little has changed in Sweden, as contemporary concerns mirror those of the Sámi during the 1980s. During the Skattefjäll (Taxed Mountain) Case in 1981 the South Sámi directly challenged the Reindeer Husbandry Act's interpretation of Indigenous rights to land, though it has continued to cause conflict over Sámi identity and land rights (Svensson, 1986). During the legal battle, Tom G. Svensson of the Arctic Institute of North America reported the opinion of one Sámi:

> As it is now the Sámi are only listened to, but when the final decisions are made, there is little to no consideration regarding the Sámi viewpoints. There are so many encroachments nowadays and life here becomes very insecure.
>
> (Svensson, 1986, pp. 212–213)

Just as the Skattefjäll Case began to set a precedent for recognizing, if simultaneously subverting, the Indigenous rights of the Sámi, the Kallak mining project represents the continued and dangerous trend of disregarding Sámi rights in favor of lucrative industrial development in the Arctic.

Neoliberal paradigm perspectives

From within the neoliberal paradigm, the proposed mining site in Gállok represents a unique opportunity to have all parties involved in the mine's construction work together for collective economic gain. On its website, Beowulf Mining reflects the basic liberal view that people will cooperate for mutual benefits. It emphasizes the importance of its local partners and incorporating diverse interests in Sweden to try and promote the mining project in Gállok to the public (Beowulf Mining, 2017). In 2016, the Beowulf CEO responded to the Sámi open letter by assuring the Sirges and Jåhkågasska Sámi "villages" that Beowulf wants to involve all stakeholders in the development of the mine, to listen to their concerns and to have the mine and reindeer herding coexist (Beowulf Mining, 2016).

While the Sámi claim Beowulf has not consulted them directly, the Beowulf CEO uses rhetoric that reflects neoliberal values of cooperation for mutual benefit. Both CEO Sinclair-Poulton and his successor Kurt Budge have discussed Beowulf's desire to work with stakeholders in Gállok and have a "dialogue" with the local Sámi communities, even though the Sámi have clearly stated they will not accept the construction of a mine in Gállok (Solly, 2016). Given that free trade is one of the main tenets of liberalism and neoliberalism promotes the expansion of global markets, from a neoliberal perspective the partnership of Beowulf Mining Plc and its Jokkmokk subsidiary represents an ideal way for Jokkmokk companies to participate in the global economy while boosting the local economy. In theory the transnational operation in Gállok would benefit all parties, but in practice the reservations of those directly affected by this enterprise are ignored in favor of transnational industrialization efforts.

Moreover, from the perspective of the Swedish state and municipal governments working within the dominant neoliberal paradigm, recognizing the Sámi's claim to the land could threaten a key attribute of the state, its sovereignty. Giving the Sámi a privileged claim to part of that territory would challenge the authority of the state. Thus, Indigenous rights claims to self-determination and natural resource management under international law are viewed by states as a threat to their authority and security (Gjelde-Bennett, 2017). Examining the Sámi's claims to Indigenous rights to self-determination from a geopolitical perspective, Dr. Oleg Kobtzeff of the American University of Paris contends that recognizing the right to self-determination "will always be perceived, either by the enthusiasts or on the contrary by the paranoid as a first step towards real, full independence and sovereignty" (Gjelde-Bennett, 2017, p. 21). The Saami Council has stated that the goal of Sámi self-determination is not to form a separate state, but to maintain autonomy over their traditional, transnational area (Sámiráddi, 2017). Though that alone makes the term self-determination problematic for realizing the Sámi's Indigenous rights. Notably when Sweden became a Universal Declaration of the Rights of Indigenous Peoples (UNDRIP) signatory in 2007, the Swedish government representative, Ulla Strom, explicitly expressed concern that the right to self-determination could pose a threat to the "territorial integrity or political unity of sovereign and independent States" (UN General Assembly, 2007). Moreover, despite constitutional recognition of the Sámi as peoples, the Saami Council reports little has changed since 2011 to effectively realize Sámi rights to self-determination (Sámiráddi, 2016).

Therefore, from a neoliberal perspective it is within the interests of both the mining company and the Swedish state to allow the mine construction based on Beowulf's promise to cooperate with local partners in Jokkmokk. There would be mutual benefits, the company claims, and the mine would bolster the local economy by opening it up to the international market. Using the land in this way would also maintain the state and municipal governments' control over that territory, avoiding any challenge to their authority. As the state is the most important actor in global politics, the Swedish state's interests in economic growth and protecting its sovereignty present formidable obstacles to ruling in favor of the Sámi communities. The legitimization of Indigenous rights domestically could alter the existing power structures that give preferential treatment to states operating within a neoliberal paradigm. Thus, states maintain ultimate authority over their territories, dictating the outcomes of conflicts within their borders.

Indigenous paradigm perspectives

Within the Indigenous paradigm, the environmental integrity of the land in Gállok should be maintained and the land accessible for reindeer husbandry and other traditional activities from an Indigenous ontological understanding of sustainability. Specifically concerning industrialization in Sápmi, human geographer Tiina Jääskeläinen (2020) contends in her literature review that the current sustainability criteria for mineral extraction projects within Sápmi fail to address the needs and concerns of the Sámi communities they would affect because the assumed concept of sustainability used for these environmental impact assessments do not take into consideration Indigenous ontologies or epistemologies that include the sociocultural aspects of environmental sustainability. Despite several international conventions that advocate for free, prior and informed consent (FPIC) and Indigenous self-determination, these key concepts are neither well defined nor consistently enforced. Moreover, government administrators typically proceed under the false assumption that Indigenous communities such as the Sámi will inevitably become a part of industrialization, and therefore a mining project will help develop the area as a whole to the benefit of all (Jääskeläinen, 2020).

The Sirges and Jåhkågasska Sámi communities have inhabited and utilized the land surrounding the proposed mining site in Gállok for reindeer herding, fishing and hunting as Sámi communities have for centuries (Gjelde-Bennett, 2017). Therefore, the Sámi communities claim the land in Gállok is where they belong and that they have a strong sense of responsibility toward it. It is also what has allowed them to continue their traditional occupation of reindeer husbandry and maintain significant aspects of their culture that revolve around the practice. Currently only about 3,000 Sámi practice traditional reindeer husbandry in Sweden, and just 900 rely on herding as a primary source of income (Williams, 2003,

Minority Rights Group International, 2012). While it is hazardous in implementing Indigenous rights to assume that all Sámi herd reindeer, as some Nordic governments already have to diminish Indigenous rights domestically, reindeer husbandry still represents an important aspect of Sámi cultural identity for many and should be protected.

The Kallak mining project thus threatens the environment and undermines Indigenous claims to their traditional lands for maintaining their Indigenous culture and identity. The Sirges and Jåhkågasska Sámi communities and the Swedish Sámi Parliament have publicly expressed concern that a mine in Gállok will irreversibly destroy the natural environment, displace entire Indigenous communities and violate international and domestic laws for protecting Sámi collective and individual rights. If Beowulf's concession is eventually turned down, it could legitimize the Sámi communities' position on natural resource management in Sweden and possibly Indigenous claims to the land in Gállok. This case could set a precedent for recognizing the rights of Sámi communities in other parts of Sweden and even throughout other Nordic countries, where there are still numerous conflicts over natural resources. However, if the concession is approved, the Swedish government will presumably continue to prioritize economic expansion over Indigenous rights and ignore domestic and international legislature which should protect Sámi communities against unwanted industrial resource extraction projects.

From within the Indigenous paradigm, the inability of the Sirges and Jåhkågasska Sámi communities to effectively block Beowulf Mining's concession represents both the unequal systems of power Sámi communities must navigate to assert their voice in political debates and the failure of systemic change due to conflicting interests. In this case study, the Swedish Sámi community clearly must work within the neoliberal political system in order to gain legitimacy for their claims. Although the proposed mine violates core values of the Indigenous paradigm, in that it negatively impacts the environment and ignores the protests of the communities it affects, Indigenous rights proponents have to argue that the mine is in breach of international and national law. Therefore, they need to focus on their individual rights, appealing to liberal and neoliberal values of universal human rights and international rule of law regulated by international institutions.

In their letter to Beowulf CEO Sinclair-Poulton in 2011, the Saami Council cites multiple national and international legal documents in their argument that the mine infringes upon the Sámi communities' basic rights, including the Swedish Constitution, the European Convention on Human Rights, the UN Convention on the Elimination of Racial Discrimination, the Universal Declaration on Human Rights, and the Universal Declaration of the Rights of Indigenous Peoples (UNDRIP) (Solly, 2012). The Sámi rely on documents which reflect liberal and neoliberal values that were created in a neoliberal-dominated international system to try and uphold the often-ignored Indigenous paradigm in global politics.

The Nordic Sámi Convention: bridging the gap

To halt Arctic industrial expansion, the Saami Council made a daring, innovative attempt to realize collective Indigenous rights transnationally by formally reasserting the Sámi are one people across state borders and therefore demand the same Indigenous rights across Nordic countries. This unique approach would theoretically allow the Sámi to break out of the dominant neoliberal paradigm to assert their own rights according to the Indigenous paradigm, which affirms the Sámi's transnational identity and intimate connection to the natural environment. Moreover, the NSC incorporates elements of the dominant neoliberal international system to gain legitimacy as a regional institution of Indigenous rights, closing the gap between international and domestic legal practices.

The Swedish government has realized some of the Sámi's collective rights under international law via the Sámi Parliament, but little has changed recently to ameliorate the Sámi's legal and political status for effectively securing their Indigenous rights. Established by the Swedish Parliament Act of 1993, Sámediggi/the Sámi Parliament is a representative organization comprised of 31 elected members with "the primary task of monitoring issues concerning the Sámi culture" (Sámediggi, 2016). In 2007, Sámediggi was given administrative control over the Swedish reindeer herding industry. However, as a government agency it is limited in its autonomy in that it must implement government policies. The Sámi Parliament reports that this often places its members in a difficult position, as they are obligated to uphold government policies which may be contradictory to Sámi interests (Sámediggi, 2016). Similarly, United Nations Special Rapporteur for the Rights of Indigenous Peoples, James Anaya, contends the Sámi Parliamentary Act in Sweden "is of particular concern" (Anaya, 2011, p. 12). According to the Swedish government, the Act was not initially "intended to be a body for Sámi self-government," which highlights the Parliament's dual character as currently limiting the Sámi's self-determination (Anaya, 2011, p. 12). The Swedish Sámi Parliament confirms on its website that for now it is "not a body for Sámi self-determination," although the Sámi possess "a right to cultural autonomy and this requires a certain degree of self-determination" (Sámediggi, 2016).

This widening gap between Indigenous rights in national versus international law continues to create conflict. The right to self-determination under international law to include greater autonomy and influence over natural recourses is glaringly absent in domestic law, yet many scholars and international law experts argue it is essential for conflict resolution (Gjelde-Bennett, 2017). In a report to the ICCPR Human Rights Committee in 2016, the Saami Council (Sámiráddi, 2016) described the situation of the Sámi in Sweden as a "kind of twilight zone where rule of law is not always guaranteed." This has created frustrations with the Swedish government, as the gap between international and domestic law places the Sámi in a "state of limbo," stalling cultural preservation and development efforts (Civil Rights Defenders, 2016, p. 10).

Defining Sámi self-determination on the transnational level

To overcome the challenges for implementing international Indigenous rights domestically, the NSC attempted to concretely define self-determination, empower the Sámi Parliaments and establish clear lines of communication with state governments to ensure the Sámi's rights to representation.

Creating a single comprehensive definition of self-determination, the NSC tried to reinforce rights given to Indigenous peoples by international law. Acknowledging the neoliberal value of international rule of law and utilizing international institutions, the NSC combined interpretations of multiple law bodies. The draft NSC reads thus:

> As a people, the Saami have the right of self-determination in accordance with the rules and provisions of international law and of this Convention. In so far as it follows from these rules and provisions, the Saami people have the right to determine their own economic, social and cultural development and to dispose, to their own benefit, over their own natural resources.
>
> (Sámediggi, 2017)

From the influential but legally non-binding 2007 UNDRIP, the NSC initially claimed the Sámi's right to self-determination under the third article so that they may "freely determine their political status and freely pursue their economic, social and cultural development" (OHCHR, 2013). Similar to the ILO Convention No. 169 that grants Indigenous peoples the right "to exercise control over their own institutions, ways of life and economic development and to maintain and develop their identities, languages and religions" (C169, 2017), the NSC recognized the importance of the natural environment to Indigenous cultural identity and tried to secure the Sámi Parliaments' control over traditional lands and natural resources.

In the dominant neoliberal paradigm, states and IGOs at the international level have historically drafted and delegated Indigenous rights. While attempts to create international law for Indigenous rights have recently progressed, sovereign states remain the most powerful actors in global politics. Thus, states can easily redefine or blatantly ignore Indigenous rights institutions such as UNDRIP. To avoid directly challenging state sovereignty, the NSC clearly defines the right to self-determination for the Sámi as greater decision-making powers, but not as the equivalent of state sovereignty.

Uniquely, the NSC not only demands the domestic recognition of Indigenous rights in the framework of international institutions created within a neoliberal paradigm, but also utilizes elements of international precedents to draft a set of unique rights for the Sámi in accordance with the Indigenous paradigm. Concerning the continued conflicts over natural resources in the Arctic, the NSC asserts that the resources within Sámi traditional territories are under Sámi control, as they concern not only reindeer husbandry but also the overall integrity of the natural environment. The fundamental relationship between Indigenous peoples and the environment within the Indigenous paradigm is thus recognized and protected. The NSC's interpretation of Sámi self-determination seeks to resolve limitations to implementing Indigenous rights based on state guidelines for Sámi identity. For instance, the 1751 Peace Treaty of Strömstad and the 1971 Sámi Reindeer Husbandry Act in Sweden specifically grant the Sámi access to natural resources for reindeer husbandry. While reindeer husbandry is an important feature of Sámi culture, not all Sámi herd reindeer, and this draft Convention thereby creates a more holistic view of both Indigenous self-determination and Sámi identity

(Gjelde-Bennett, 2017). Sámi would not have to be engaged in herding in order to exercise their Indigenous rights, and the non-herding majority could also influence natural resource management. A greater legal ownership of traditional Sámi lands could therefore prevent industrial development (like the proposed Kallak mining project) which is harmful to the Sámi's own development and the natural environment on behalf of all those living there.

Moreover, the NSC aims to overcome the challenges of the limited powers of the Sámi Parliaments in Norway, Sweden and Finland and the lack of cooperation between the Sámi and national governments by giving the Sámi the right to representation on the international and domestic levels. The draft NSC empowers the Sámi Parliaments to make "independent decisions on all matters where they have the mandate to do so under national or international law," and the right to consultation with the state governments on all policies concerning them (Koivurova, 2006, p. 118). Similar to the right to consultation given in the ILO 169, the draft NSC gives the Sámi the right to negotiate all matters of importance to the Sámi. Moreover, no government can pass any measures which could "damage the basic conditions for Saami culture, Saami livelihoods or society" (Sámediggi, 2017). Though still referencing national and international law, the establishment of specific procedures on policies that directly affect Sámi communities could be crucial for ensuring governments have provided the Sámi with the political and moral authority necessary for resolving reoccurring problems on land rights and natural resource management. Once established, these structures could create more effective lines of communication for conflict resolution.

However, the NSC in its latest form does not grant the Sámi rights to self-determination and necessary decision-making powers to block harmful industrial projects in the Arctic. In practice, the right to consultation alone, which some Sámi Parliaments already technically possess, is not only insufficient but also frequently ignored by states. The Sámi Parliaments must be granted greater decision-making powers beyond rights to consultation in order to effectively negotiate and cooperate with powerful state governments and other non-state actors.

In this sense, the acceptance of greater Sámi self-determination is not that they can establish a new sovereign state, but rather to determine a new, more equitable relationship between state governments, industries and Indigenous communities. The Saami Council and the Sámi Parliaments desire greater involvement in state decision-making processes to influence policies which directly affect them (Sámiráđđi, 2016). The draft NSC thereby promotes constructive cooperation between the Indigenous and non-Indigenous bodies by guaranteeing Sámi the right "to be represented in public councils and committees and in intergovernmental meetings when these deal with matters that concern the interests of the Saami" (Koivurova, 2006, p. 118). Whether in a city committee or in an intergovernmental forum, the Sámi must have greater representation in the public political domain to promote their interests and secure their Indigenous rights. By being more involved in the public political sphere, the Sámi would gain invaluable influence and legitimacy on the international, state and local levels, balancing the representation of the neoliberal and Indigenous paradigms in Indigenous rights domestic implementation.

If this more equitable system were in place, the establishment of a mine on traditional Sámi reindeer herding territory would have automatically involved the

Sámi in the decision-making process for the prospective Beowulf mine in Gállok. The distinction made by the Saami Council on Sámi self-determination would have been essential for redefining Indigenous rights to self-determination so that state governments do not perceive them as a direct threat to state sovereignty.

The failings of the Nordic Sámi Convention

Despite high hopes, some Sámi and other Indigenous scholars have criticized the latest version of the NSC released in 2017 for being a colonial tool, as Indigenous claims to rights have been diluted or revoked through their interpretations by the Nordic state governments. Sámi scholar Magne Ove Varsi writes in his article "A disappointing Nordic Sámi Convention" that after much anticipation the final version of the NSC lacks real ambition to advance Indigenous rights beyond the status quo and alarmingly reduces Sámi self-determination to the problematic right to consultation (Varsi, 2017). Rauna Kuokkanen of the University of Lapland declares that the final draft of the NSC will, "seal the colonization of the Sámi through a devious legal document that claims to strengthen the status and rights of the Sámi as an Indigenous people but in reality does the very opposite" (Kuokkanen, 2017).

Kuokkanen (2017) argues the NSC reproduces pre-existing interpretations of Indigenous rights as being limited to cultural rights:

> The states don't mind if we speak our languages and wear our *gáktis* in our festivals and cultural activities. It fits well with the intentions of the neoliberal state of displaying and commodifying Indigenous peoples and their traditions – in fact, one of the articles (Art. 35) recognizes the commodity value of Sámi culture, history and nature for creative industries and tourism. The provision is a travesty of Indigenous rights.

Considering Indigenous scholars' criticisms of the NSC within the broader context of paradigm conflicts in the Nordic countries, arguably the NSC failed because of the larger, unequal power structures within which the document was produced. The NSC initially reflected the Indigenous paradigm for pursuing Indigenous rights, a feat within itself. However, the NSC was altered during negotiations to make it conform to the neoliberal paradigm. Instead of helping the Sámi realize essential international rights to self-determination and natural resources to halt industrial expansion into the Arctic, the document reproduced the same system of inequality and exploitation. Viewing the NSC as the site of a paradigm conflict explains the shortcomings of the Convention; the NSC was formed and negotiated within a greater system of inequality based in colonial history that favored the interests of more powerful state governments operating within a neoliberal paradigm. Historically linked to territory and security, state sovereignty is consistently prioritized over Indigenous claims to traditional lands and self-determination (Gjelde-Bennett, 2017).

The NSC first sought greater cooperation between state governments and the Sámi Parliaments in order to minimize the effects of state borders on the Sámi people for realizing their Indigenous rights. However, negotiations within a neoliberal-dominated system became another repetitive standoff between states and Indigenous peoples, state sovereignty versus self-determination, making the negotiations a zero-sum game. What the Sámi gain is what states lose and vice versa. The state actors involved in the NSC negotiations clearly did not move beyond perceiving the nature of the current international system from outside of the neoliberal paradigm.

The Saami Council tried to deviate from the prescribed script of the neoliberal versus Indigenous paradigms by asserting their own Indigenous rights in the Nordic countries with components of both paradigms in the Nordic Sámi Convention. However, the more powerful state governments operating within the dominant neoliberal paradigm manipulated the NSC so that it reproduced the current system of inequality, and some Sámi passionately argue made it even more unequal.

The current situation in *Gállok* and beyond

Returning to the case study in *Gállok* and how the situation has progressed since the latest version of the NSC was released, there remains an urgent need for creative solutions to resolve paradigm conflicts in the Arctic. From 2018, Beowulf's latest public update on the Kallak Iron Ore Project (as it is now referred to) states the company's exploitation concession continues to be postponed by the Swedish government (Beowulf Mining, 2018). The Kallak project page on the Beowulf Mining website features a 2017 study conducted by Copenhagen Economics, titled "Kallak—A real asset, and a real opportunity to transform Jokkmokk" (Beowulf Mining, 2020). The study summary reflects multiple neoliberal values in highlighting the shared economic benefits and by saying that Beowulf's proposed Kallak mine shall potentially create 250 direct jobs over 25 years. It would ensure that "Jokkmokk maximises the benefits it receives" (Beowulf Mining, 2020).

In a 2018 video on the current state of the Kallak Iron Ore Project from Beowulf's corporate webpage, the mayor of Jokkmokk Robert Bernhardsson concurs, "this investment will benefit Jokkmokk, and generate many jobs, and also great hopes for the future" (Beowulf Mining, 2018). The video goes on to explain that the project delays are due to Swedish political conflicts involving the Green Party, "environmental demands all met by the company" and "Sámi people claiming their Indigenous rights" (Beowulf Mining, 2018). Both current Beowulf CEO Kurt Budge and Jokkmokk Mayor Bernhardsson state that the solution to dealing with these issues is by opening a dialogue. Budge contends, "we will do our utmost to demonstrate that we want to work in partnership, not just with the Sámi, but with all sections of the community" (Beowulf Mining, 2018). No indication is given of how the company proposes to accomplish this goal. Bernhardsson merely states that neighboring municipalities have come to agreements on "similar issues" (Beowulf Mining, 2018). Evidently, Beowulf's overall rhetoric and approach to mitigating controversy does not appear to have changed significantly for nearly ten years.

Meanwhile the surrounding Sámi communities remain starkly opposed to the mining project. In 2019, Truls Anderssen's documentary principally follows Tor Lundberg and his family who live in Gállok. A member of the local Sámi community, Tor discusses hearing about the mining projects in northern Sweden on the radio, demonstrating the disconnect between industry and the Indigenous communities. This is particularly apparent when Tor quotes the mining companies using the name "Kallak," the Swedish name for his home, which he usually refers to as "Gállok" in his Sámi language. Contrary to the neoliberal mindset that the Sámi are inhibiting progress by refusing to utilize the natural resources in Gállok, Tor offers a local Indigenous perspective on the value of the natural environment:

> The Sámi have always known about the iron ore here because this mountain, Atjek, is often struck by lightning. Atjek means thunder in the Lule Sámi language, the Mountain of the Thunder God, a sacred place.
>
> (*Gállok*, 2019)

In these few lines, Tor reveals the local Sámi are not ignorant of the value of the iron ore deposits, as some proponents of the mining project submit. The Sámi simply perceive it differently within their Indigenous paradigm, valuing their relationship to the land and what that means to their collective history and culture.

One of the most jarring scenes of the film involves Tor and his daughter Astrid discussing the disastrous consequences to the natural environment and their local community should Beowulf be allowed to proceed with the mine. Tor takes Astrid to participate in a municipal meeting in Jokkmokk concerning the potential long-term effects of the proposed mine in Gállok. During the debate, he says:

> Of course, it's easy to sit down south and think mines are great. But they affect us who live here, our reindeers, our tourism and all who live on what nature provides. Food production is more important than anything up here, particularly fishing.
>
> (*Gállok*, 2019)

Anderssen also highlights the discrimination Tor's family faces daily. During one particular trip to the supermarket, other shoppers comment to the family that the Sámi are "begrudging other people getting work" in response to their daughter wearing *gákti*, traditional Sámi clothing, in the store (*Gállok*, 2019). Nevertheless, Tor remains opposed to Beowulf's mining project, maintaining an indigenist perspective on sustainability. He repeatedly emphasizes how the destructive aftereffects of mining on the environment are permanent while the economic benefits are only temporary.

Unfortunately, the ongoing Kallak mining controversy is not an isolated incident but is representative of industrial projects and the resulting paradigm conflicts across the Nordic countries. In 2019, the Norwegian government approved a mining concession for a copper mine to be constructed in Repparfjord that will not only disrupt Sámi reindeer herding and fishing industries, but will also irreparably damage the natural environment (Nilsen, 2019). The Nussir mining company's CEO Vidar Rune Late's comments echo those of Beowulf's CEO as he emphasizes the new job opportunities the mine would provide and how investing millions of dollars into the project would bolster the local municipality's economy (Pietromarchi, 2019). Similar to the situation in Gállok, local activists and Sámi communities in Repparfjord have demonstrated against the mining project, concerned about its long-term environmental, economic and social impact. Erik

Reinert, former professor of reindeer economics at Sámi allaskuvla/Sámi University of Applied Sciences reports:

> For hundreds, jobs that will exist at best for a few decades, the government is ruining a fjord, reindeer summer pastures and fisheries … food production for hundreds of years ahead, if not forever.
>
> (Pietromarchi, 2019)

Unlike in Gállok, the Norwegian government has already approved Nussir's concession, and those against the decision are rightly concerned that allowing this copper mine's construction will impact future governmental decisions and encourage the mining industry in other parts of Sápmi (Pietromarchi, 2019).

Conclusion: possibilities for compromise?

In summation, an analysis of the mining controversy in Gállok through the neoliberal and Indigenous paradigms exposes the dominance of the neoliberal paradigm in the interactions between state governments, industry and Indigenous peoples. As a result, Indigenous peoples who hold an Indigenous worldview are at a disadvantage, as they have to utilize neoliberal tools, which uphold the ultimate authority of the state to justify claims to Indigenous rights that could potentially challenge state sovereignty. Finding common ground between the neoliberal and Indigenous paradigms would therefore be the key to forming new institutions to realize international Indigenous rights domestically. The Saami Council's proposed NSC represents a historic transnational, subregional institution which had the potential to guarantee the same Indigenous rights for all Sámi in Norway, Sweden and Finland. The NSC is especially unique because it was initially created by Indigenous peoples for Indigenous peoples, rather than by the usual neoliberal institutions implementing policies for Indigenous peoples.

Though flawed in its final draft version, the initial spirit of the NSC represents compelling evidence of norms and practices within the international system beginning to change in order to recognize both neoliberal and Indigenous paradigms and to empower Indigenous peoples for realizing their own collective rights. While hope for achieving this reality in its totality may appear remote within the near future, it would be remiss to deny that the current system has not been the product of historic processes and hazardous to assume that it will not continue to change through ongoing interaction over the coming decades. However, it should not be denied that the systematic discrimination of Indigenous peoples and exploitation of natural resources on their traditional lands has been a long-established feature of the international system. Indigenous scholar Victoria Tauli-Corpus contends these two worldviews are irreconcilable, "There is an inherent tension between the claims of nation-states to eminent domain and national sovereignty, and those of Indigenous peoples who assert their rights to their traditional lands" (Tauli-Corpuz, 2006, p. 216). While it is true that the conceptual bases of these paradigms have been at odds in practice for decades, especially concerning Indigenous rights to self-determination and their traditional lands, the NSC's existence as a proposed transnational convention for realizing Indigenous rights suggests that it might be possible to reconcile these differing worldviews and that a more equitable international system could be formed in future.

There has already been a shift in global politics led by Indigenous scholars and activists to empower Indigenous peoples globally and revalue Indigenous perspectives and methodologies within the international system. This is clearly demonstrated by the legal victory of the Māori Whanganui iwi in Aotearoa/ New Zealand, successfully protecting Te Awa Tupa/the Whanganui River as an ancestor by granting it individual rights as a living entity in 2017. This new legal precedent incorporates elements of both paradigms to resolve a 140-year conflict, as the river is protected using Indigenous logic by equating its worth to that of a human being and it is legally enforced within a neoliberal individual rights framework (Roy, 2017).

Although these recent developments do not mean that Indigenous rights proponents can wait for this systemic change to happen, they must be persistent and prepared to be the vehicle for that change. Indigenous peoples like the Sámi and their allies have known this for over a century, effectively fighting for their own rights. Still, there needs to be a heightened awareness of the greater conceptual inequalities which foster exploitation and conflict between industry, state governments and Indigenous peoples. Differing worldviews must be reconciled in order to achieve more holistic long-term solutions for realizing Indigenous rights.

References

Anaya, J. (2011) *The situation of the Sami people in the Sápmi region of Norway, Sweden and Finland* [Online]. The United Nations. Available: http://unsr.jamesanaya.org/country-reports/the-situation-of-the-sami-people-in-the-sapmi-region-of-norway-sweden-and-finland-2011 (Accessed May 28, 2020).

Gállok: The Battle for Sami Rights in Sweden 2019. Directed by Anderssen, T. Sweden: Deep Sea Reporter.

Baeten, G., Lund Hansen, A. and Berg, L. D. (2016) Neoliberalism and Post-Welfare Nordic States in Transition, *Geografiska Annaler*, 97, pp. 209–212.

Beowulf Mining (2016) *Open Letter to the Chairmen of Jåhkågasska and Sirges Sami Villages* [Online]. Beowulf Mining plc. Available: http://beowulfmining.com/news/open-letter-to-the-chairmen-of-jahkagasska-and-sirges-sami-villages/ (Accessed April 21, 2017).

Beowulf Mining (2017) *Strategy* [Online]. Beowulf Mining plc. Available: http://beowulfmining.com/about-us/strategy/ (Accessed April 21, 2017)

Beowulf Mining (2018) *Kurt Budge CEO Beowulf Mining plc Kallak Iron Ore Project Interview* [Online]. Beowulf Mining plc. Available: https://beowulfmining.com/kurt-budge-ceo-beowulf-mining-plc-kallak-iron-ore-project-interview/ (Accessed March 17, 2020).

Beowulf Mining (2020) *Kallak- Iron Ore* [Online]. Beowulf Mining plc. Available: https://beowulfmining.com/projects/sweden/kallak/ (Accessed May 28, 2020).

Boland, H. (2016), *Beowulf Mining's Concession Faces Extra Environmental Impact Criteria* [Online]. London South East. Available: http://www.lse.co.uk (Accessed April 15, 2016).

Civil Rights Defenders (2016) *Sweden's compliance with the International Covenant on Civil and Political Rights (ICCPR)* [Online]. OHCHR. Available: https://tbinternet.ohchr.org/Treaties/CCPR/Shared%20Documents/SWE/INT_CCPR_CSS_SWE_23074_E.pdf (Accessed May 30, 2020).

Forss, A. (2015) *A New Era of Exploitation? Mining Sami Lands in Sweden* [Online]. Cultural Survival. Available: https://www.culturalsurvival.org/publications/cultural-survival-quarterly/new-era-exploitation-mining-sami-lands-sweden (Accessed April 14, 2020).

Gjelde-Bennett, K. (2017) *Challenges to Implementing Indigenous Rights in Scandinavia: A Multidisciplinary Analysis of the Nordic Sámi Convention*. Tacoma, Washington: Pacific Lutheran University.

Gjelde-Bennett, K. N. (2020) *Indigenous Efflorescence and Tjåenieh in Southern Saepmie: Rethinking Language Revitalization Research in Conversation with a Saemie Illustrator*. Master of Philosophy in Indigenous Studies Master, UiT The Arctic University of Norway.

Harris, L. D. and Wasilewski, J. (2004) "Indigeneity, an Alternative Worldview: Four R's (Relationship, Responsibility, Reciprocity, Redistribution) vs. Two P's (Power and Profit). Sharing the Journey Towards Conscious Evolution," *Systems Research and Behavioral Science*, 21, pp. 489–503.

C169 *C169 - Indigenous and Tribal Peoples Convention, 1989 (No. 169)* [Online]. International Labour Organization (2017). Available: https://www.ilo.org/dyn/normlex/en/f?p=NORMLEXPUB:12100:0::NO::P12100_ILO_CODE:C169 (Accessed May 29, 2020)

Jääskeläinen, T. (2020) "The Sámi reindeer herders' conceptualizations of sustainability in the permitting of mineral extraction – contradictions related to sustainability criteria," *Current Opinion in Environmental Sustainability*, 43, pp. 49–57.

Keohane, R. O. (1984) *From After Hegemony: Cooperation and Discord in the World Political Economy*, Princeton: Princeton University Press.

Koivurova, T. (2006) "The Draft for a Nordic Saami Convention," *European Yearbook of Minority Issues*, 6, pp. 103–136.

Kovach, M. (2010) *Indigenous Methodologies - Characteristics, Conversations, and Contexts*. Buffalo: University of Toronto Press.

Kuokkanen, R. (2017) Smoke and Mirrors of the Sami Convention. Available from: http://rauna.net/2017/01/29/smoke-and-mirrors-of-the-sami-convention/ (Accessed January 29, 2020).

Lightfoot, S.R. (2016) *Global Indigenous Politics: A Subtle Revolution*. London: Routledge.

Mander, J. (2006) "Introduction: Globalization and the Assault on Indigenous Resources," in J. Mander and V. Tauli-Corpuz (eds.) *Paradigm Wars: Indigenous Peoples' Resistance to Globalization*. San Francisco: Sierra Club Books, pp. 3–10.

Minority Rights Group International (2012) *State of the World's Minorities and Indigenous Peoples 2012 - Case study: Sami rights to culture and natural resources* [Online]. refworld. Available: https://www.refworld.org/docid/4fedb3de37.html (Accessed May 30, 2020).

Nilsen, T. (2019) *Norway greenlights copper mine with tailings to be dumped in Arctic fjord* [Online]. The Barents Observer. Available: https://thebarentsobserver.com/en/industry-and-energy/2019/11/norway-greenlights-copper-mine-tailings-dump-arctic-fjord (Accessed April 14, 2020).

Pantzare Information Ab (2010) *Facts about Norbotten* [Online]. County Administrative Board of Norbotten. Available: http://www.lansstyrelsen.se (Accessed April 14, 2017).

Pietromarchi, V. (2019) *Norway approves disputed Arctic copper mine despite local protest* [Online]. Aljazeera. Available: https://www.aljazeera.com/news/2019/02/norway-approves-disputed-arctic-copper-local-protest-190214170223061.html (Accessed April 14, 2020).

Roche, G., Virdi Kroik, Å. & Maruyama, H. (2018) *Indigenous Efflorescence*. Canberra: ANU Press.

Roy, E. A. (2017) New Zealand river granted same legal rights as human being [Online]. *The Guardian*. Available: https://www.theguardian.com/world/2017/mar/16/new-zealand-river-granted-same-legal-rights-as-human-being (Accessed March 17, 2017).

Sámediggi (2016) *Sámediggi* [Online]. Sámediggi. Available: https://www.sametinget.se (Accessed May 29, 2020).

Sámediggi (2017) *Nordic Saami Convention* [Online]. Sámediggi. Available: https://www.sametinget.se/105173 (Accessed December 29, 2020).

Sámiráđđi (2016) *Sweden's compliance with the International Covenant on Civil and Political Rights (ICCPR)*. Briefing paper submitted for the UN Human Rights Committee's review of Sweden during its 16 session, 7-13 March 2016.

Sámiráđđi 2017. *Tråante Declaration* [Online]. Available: https://static1.squarespace.com/static/5dfb35a66f00d54ab0729b75/t/5e722293aee185235a084d70/1584538266490/TRA%CC%8AANTE_DECLARATION_english.pdf (Accessed January 9, 2021).

Solly, R. (2011) *British Beowulf caught drilling illegally* [Online]. Available: https://londonminingnetwork.org/2011/12/british-beowulf-caught-drilling-illegally/ (Accessed May 29, 2020).

Solly, R. (2012) *British Beowulf reported again for breach of Swedish Minerals Act* [Online]. London Mining Network. Available: https://londonminingnetwork.org/2012/01/british-beowulf-reported-again-for-breach-of-swedish-minerals-act/ (Accessed May 29, 2020).

Solly, R. (2016) *Beowulf wants dialogue with Sami but won't take no for an answer!* [Online]. London Mining Network. Available: https://londonminingnetwork.org/2016/02/beowulf-wants-dialogue-with-sami-but-wont-take-no-for-an-answer/ (Accessed May 29, 2020).

Svensson, T. G. (1986) "Ethnopolitics among the Sámi in Scandinavia: A Basic Strategy toward Local Autonomy," *Arctic*, 39, pp. 208–215.

Tauli-Corpuz, V. (2006) "The Prospect Ahead," in J. Mander, and V. Tauli-Corpuz (eds) *Paradigm Wars: Indigenous Peoples' Resistance to Globalization*. San Francisco: Sierra Club Books, p. 216.

United Nations General Assembly (2007) "General Assembly Adopts Declaration on Rights of Indigenous Peoples; 'Major Step Forward' Towards Human Rights for All, Says President," *United Nations* [Online]. Available: https://www.un.org/press/en/2007/ga10612.doc.htm (Accessed May 29, 2020)

United Nations Human Rights Office of the High Commissioner (OHCHR) (2013) *Indigenous Peoples and the United Nations Human Rights System* [Online]. Available: https://www.ohchr.org/Documents/Publications/fs9Rev.2.pdf (Accessed May 29, 2020).

Varsi, M. O. (2017) Kommentar: En skuffende nordisk samekonvensjon. *NRK Sápmi*. Available: https://www.nrk.no/sapmi/kommentar_-en-skuffende-nordisk-samekonvensjon-1.13330811 (Accessed May 29, 2020).

Virtanen, P.K., Siragusa, L. & Guttorm, H. (2020) "Introduction: toward more inclusive definitions of sustainability," *Current Opinion in Environmental Sustainability*, 43, pp. 77–82.

Williams, S.M. (2003) "Tradition and Change in the Sub-Arctic: Sámi Reindeer Herding in the Modern Era," *Scandinavian Studies*, 75, pp. 229–256.

Wilson, S. (2008) *Research is Ceremony: Indigenous Research Methods*. Halifax: Fernwood publishing.

10 Revisiting the governance triangle in the Arctic and beyond

Monica Tennberg, Else Grete Broderstad and Hans-Kristian Hernes

From the governance triangle to meta-governance

Large-scale projects to extract energy resources, minerals and fish are attractive to governments as well as for local communities. They promise to bring income, employment and well-being, while concerns over social and environmental consequences of such projects are also widely known and shared. Our cases in this book—of wind power development, aquaculture and mining—represent extractive industries. Recent Arctic research has focused on the conflicts between extractivism, Indigenous self-determination and government policies (Kuokkanen, 2019; Lawrence and Moritz, 2019; Willow, 2019; Alcantara and Morden, 2019; and Tysiachniouk, Petrov and Gassiy, 2020). Extractivism pertains to the industrial use of natural resources and land: it is a response to ever-growing global resource demands and has become increasingly dominated by foreign investments, privatization of industrial activities and company-led practices of corporate social responsibility as "neo-extractivism" (Wilson and Stammler, 2016; Junka-Aikio and Cortes-Severino, 2017).

Governance of natural resources refers to the principles, institutions and processes that determine how power, obligations and responsibilities over natural resources are exercised, how decisions are taken, and how peoples and their communities participate in, benefit from and oppose the extraction of natural resources. The principle of self-determination is central in natural resource governance for Indigenous peoples and their political, social and economic rights confirmed in numerous global human rights conventions and declarations discussed in this book. These rights have been applied widely and differently in different national contexts, but as the Australian sociologist Louisa Humpage (2010, p. 539) notes, "uncertainty thus remains about the best mix of recognition and redistribution needed to produce good outcomes for Indigenous peoples in terms of both welfare *and* greater Indigenous autonomy and control."

Central to governance is the way governments, authorities, private bodies and non-governmental organizations interact with each other while aiming to solve governance challenges, avoid failures and create opportunities for better governing. Our approach seeks to go beyond governance as practical problem-solving or general rule-making, and rather view it as meta-governance (Kooiman and Jentoft, 2009; see also Meuleman, 2008; Jessop, 2011). Meta-governance embraces

DOI: 10.4324/9781003131274-10

principles, norms and values: it is about the normative bases that underpin different forms of natural resource governance and also includes the application of principles guiding interactions and communication between agencies and responsible institutions (Kooiman et al., 2005).

The different modes of governance here refer to *hierarchical, state-led governance* with authority and legitimacy as its main values; *industries-led market governance*, where the main values are profit, effectiveness and time; and *civil society-based network governance*, which is steered most of all by trust and consensus. These modes usually appear in various combinations and together they also produce different kinds of interactions between different parties and governance failures (Meuleman, 2008, 2019). From the meta-governance perspective, governance failures stem from the mix of different modes of governance. Governance failures can be expected, firstly, if there is an institutional mismatch between the issues to be governed and institutional arrangements, and secondly, when capacity and resources for governance are lacking. (See also Smith, 2008; Larsen and Raitio, 2019; Meuleman, 2019; La Cour and Aakerstroem Andersen, 2016).

Our analysis of meta-governance centers on how Indigenous self-determination as a major, internationally recognized principle and nationally implemented norm is interpreted in interactions between Indigenous communities, states and extractive industries. In these interactions, three principal sites have been identified on the basis of the investigated cases. First of all, there are the legal processes balancing economic interests, Indigenous rights and national and international commitments, which in practice become tangible in various consultation processes, authorities' decisions, court hearings and legal developments. Second, there are the different forms and procedures in private agreement-making between industry representatives and Indigenous peoples' organizations, while the third site is that of individual projects, local debates between Indigenous activists, local and national decision-makers and authorities and company representatives. It is here that the different modes of governance, values, principles and norms meet and mix, resulting in both successes and failures in natural resource governance from the Indigenous peoples' perspectives.

Our analysis examines both structural constraints and discursive opportunities that Indigenous peoples have in communicating their concerns about current extractive industry plans and projects. We have analyzed how the agency of Indigenous peoples manifests—as rights-holders, stakeholders and contesting the norms and the different mixtures in practice—in natural resource governance in Nordic countries, Russia, Canada, Australia and New Zealand. Our cases represent both positive and negative outcomes from the Indigenous peoples' perspectives.

Balancing Indigenous rights, economic interests and national commitments in court

In our cases, hierarchical governance is a combination of the countries' colonial pasts and current, mostly neoliberal, governmental policies, framing in different ways how the states apply global human rights mechanisms. From the perspective of hierarchical governance, Indigenous peoples are rights-holders. The results from our case studies show that governments are struggling to recognize this role for

Indigenous peoples. In her chapter, Else Grete Broderstad discusses state compliance with international law obligations, especially the role of article 27 of the International Covenant on Civil and Political Rights (ICCPR) in the Sámi rights development in Norway. The discussion focuses on a specific case where the Norwegian Ministry of Petroleum and Energy halted the plan for building a wind power station on traditional reindeer herding lands in 2016. The ministerial decision to revoke the existing license can be interpreted as avoiding violation of human rights through government action. However, the government's assessment of the very same article 27 has in other cases led to a different outcome. In order to understand this mismatch, Broderstad identifies different indicators for state compliance, including structural indicators to assess states' commitment to the protection of human rights in domestic legislation while process indicators evaluate their implementation. Outcome indicators are important in the assessment in concrete cases of state failure to comply with human rights. Despite these inconsistencies in state compliance with international law in Norway, Broderstad concludes that "[t]he state nevertheless remains the primary duty-bearer of human rights obligations."

The failure to honor international and national commitments is evident in the Swedish wind power cases discussed by Dorothée Cambou, Per Sandström, Anna Skarin and Emma Borg in this volume. The increasing number of court cases in the governance of natural resources is testimony to the opaque application of the central principles of the governance process, including such issues as sustainable development and Indigenous rights, and their legitimacy. The courts have become the last arena where the different economic interests, Indigenous rights and national commitments are mediated. However, due to Swedish legislation, the courts' role as mediator is limited in ensuring the protection of the rights of the Sámi. According to Cambou et al., in these Swedish cases, the final court decisions favor a market-oriented perspective of sustainable development in allowing wind energy development in the reindeer herding areas. The courts have not confronted the political imbalance between national environmental and economic interests and those of the Sámi as an Indigenous people and the sustainability of their traditional lands and livelihoods at the local level. As Cambou et al. argue, these Swedish cases "epitomize the persistent challenges faced by the Swedish courts to ensure sustainability at every level amid increasing demands to promote sustainable development for all."

Unclear obligations on company consultations with Indigenous peoples emphasizes the role of the courts, as Gabrielle Slowey shows in her chapter about a mining case in Ontario, Canada and the concerns of Ontario First Nations. As the Canadian courts now require that Indigenous nations be consulted, the state encourages and may in some cases demand that companies negotiate impact benefit agreements (IBAs) with communities as proof of such consultation before issuing the companies with permits and licenses. However, recent developments in Canadian legislation may undermine this obligation: in 2020, the Ford government in Ontario introduced a new bill (Bill 197: The COVID-19 Economic Recovery Act), which among other things modifies the main parts of the provincial environmental assessment regulations. Slowey sees this governmental move as an attempt in the shadow of the global pandemic to "further push mining and northern development onto the backs of First Nations." Slowey points out that although Ontario First Nations have constitutionally protected treaty rights and the courts have mandated consultation in advance of development projects, the Ford legislation streamlines and, in some cases, removes the established procedures and protections which are essential in areas where First Nations remain directly tied to state action and are subject to extraction. This is especially challenging given the small number of modern, post-1975 treaty land claims with accompanying self-government agreements which offer more agency for First Nations. Slowey stresses that "[w]here First Nations do not have modern treaties, there is no level playing field" between Indigenous peoples and industry plans.

The role of the state in ensuring Indigenous rights in agreement-making

In market governance, contracts are key and the role of the state is similarly important in drawing up fair agreements. In Australia, under neoliberalism, the state encourages private agreement-making for natural resource governance, which renders the industries and Indigenous peoples responsible for making such deals. Similar processes are underway in Canada, where Indigenous participation in decision-making over extractive projects on their lands is channeled through highly regulated impact assessment procedures and the negotiation of agreements with developers. The contractual nature of market governance has resulted in company-based tools for natural resource governance, such as impact benefit agreements (IBA); free, prior and informed consent (FPIC); corporate social responsibility (CSR); and social license to operate (SLO). These practices promise Indigenous communities compensation for cooperation with companies and access to their traditional areas, and construct Indigenous peoples as economically driven stakeholders and rational actors following a neoliberal governmental logic. These company-led approaches have been criticized for privatization of consultation, naturalizing market-based solutions and limiting access to important political and legal channels (MacDonald, 2011; Strakosch, 2015).

The Canadian cases in this book show the importance of state support in agreement-making between industries and Indigenous peoples. The Tlicho of Mackenzie Valley are a Canadian First Nation that have both a modern land claims treaty and self-government. The creation of the Tlicho government with law-making authority over citizens, communities and lands in the Northwest Territories has produced an Indigenous governing body which is influential as a decision-maker and resource manager in its traditional area. Horatio Sam-Aggrey discusses in his chapter the role of the impact benefit agreement, negotiated between the mining industry and Tlicho communities. The case also shows that comprehensive land claims agreements (CLCAs) in northern Canada have provided a legal framework to ensure Indigenous participation, political leverage in natural resource governance and clarity in terms of land ownership and Indigenous rights. Sam-Aggrey concludes that "it is not far-fetched to argue that in the case of the Tlicho, the relationship between the State and the Tlicho is the most important angle of the governance triangle." It has given the Tlicho political leverage in their relationship with industry, thereby slightly leveling the playing field between the two parties.

In the case of Australia, Catherine Howlett and Rebecca Lawrence investigate the agreement-making tools to support the idea that Indigenous peoples agree with and consent to resource developments on their lands and territories at the expense of their rights. The recent case of the agreement-making process for the Adani Carmichael coal mine in Queensland illustrates the problems in the agreement-making process. The Wangan and Jagalingou peoples have previously rejected the company's development proposals for the Carmichael mine. Despite the resistance, several mining leases have been issued to the company by the Queensland government without the consent of all of the Wangan and Jagalingou native title claimants. Under the current Australian legislative framework, it is not necessary to obtain consent of all Indigenous parties whose native title rights will be affected by development. Howlett points out that the Australian agreement-making practice allows only an extremely limited form of "agency," one where Indigenous peoples are forced to engage with, and consent to, tools that ultimately dispossess them.

Comparing the Nordic and Australian circumstances, Howlett and Lawrence argue in their chapter that the Nordic states are not immune to neoliberal natural resource policies; in fact, if anything, "they have in some ways been all the more willing to take on neoliberal logics and practices." Benefit-sharing and part-ownership agreements with local municipalities, landowners and neighboring (non-reindeer herding) local communities have been negotiated for wind power projects in Sweden while the Sámi have been completely marginalized in these processes. In the case of the Stekenjokk agreement between the County Board and the wind power company, the local Sámi community was not party to the agreement nor to any of the negotiations. What the Swedish state did instead, according to Howlett and Lawrence, was play the role of paternal protector by claiming to represent Sámi interests, but it also took "the role of market actor by staking a claim to a market share of any profits." In this way, the state leveled the playing field between wind power developers and landowners.

Indigenous agency in network governance

Network governance refers to situations where the state engages with non-governmental actors, such as Indigenous peoples and industries, while maintaining some degree of control over the activity of such governance networks (Jessop, 1998; Sørensen and Torfing, 2007). In our analyses, network governance has been understood as a mostly locally constituted but often rather dispersed network of representatives of Indigenous peoples, authorities from different levels of administration, local people and company representatives interacting upon a project idea or a more concrete plan. This is the least defined site of governance among our cases, and complex power relations between different actors are also at play here. Most importantly in this context, the normative power of Indigenous peoples often translates to an issue of knowledge. Knowledge and the use of various kinds of knowledge by different knowledge holders are key to creating a common understanding of the values, principles and norms to be negotiated and implemented. This is an important aspect of Indigenous agency in natural resource governance, as it entails the very ability of Indigenous peoples to participate in the planning of extractive projects, influence development in their areas and represent their interests and rights (Brattland et al.; see also Cambou et al. in this volume).

As Camilla Brattland, Else Grete Broderstad and Catherine Howlett stress in their chapter, Indigenous peoples' political agency is a result of multiple power relations that define their scope and forms of political action. The comparison between the Norwegian and New Zealand salmon industry highlights structural and discursive constraints on Indigenous agency. In both countries, Indigenous peoples are heavily involved in marine livelihoods, but in markedly different ways. Whereas the Māori hold a share of marine industry development in New Zealand, in Norway marine development is seen as the domain of the municipalities. Likewise, the process of consultation and participation in aquaculture licensing differs greatly between the two countries. The standards set by global human rights mechanisms on consultations and agreements with Indigenous local communities have been implemented in the New Zealand governance system, which serves to strengthen Indigenous agency as seen in the Marlborough Sounds case. In contrast, the lack of state recognition of Sámi presence and interests in the coastal areas of Norway, as is evident in the Vedbotn aquaculture case, seriously hinders Indigenous agency in local aquaculture development and beyond. In their analysis in this volume, Brattland and colleagues identify a clear need to recognize the obligation of local communities to consult the Sámi on issues that affect Sámi interests and livelihoods.

Networks also function in the context of coercive, authoritarian modes of governance (Kropp and Schuhmann, 2016; see also Berg-Nordlie, 2018), as becomes clear in the analysis by Marina Peeters Goloviznina in this volume. Goloviznina discusses a Russian case by applying the principle of free, prior and informed consent in her chapter about normative conflicts between Indigenous peoples and a gold mining company in the Nezhda mining plan in the Tomponskyi municipal district, northeast Yakutia, Russia. Goloviznina notes that the interactions between the extractive company, Indigenous peoples and authorities take place in the context of the rights-incompatible Russian state where the authorities deliberately replace community (Indigenous) voices and speak on their behalf. This mode of interactions encourages companies to deal with government representatives instead of working with Indigenous peoples directly. In this case, the two regional institutionalized practices in the Sakha Republic (Yakutia)—ethnological expertise and the Ombudsman for Indigenous Peoples' Rights—complement and enhance each other's work, and offer Indigenous peoples ways to broaden their participation in the local governance of natural resources. In addition, as Goloviznina points out in her chapter, the local community benefits from its networks with

The brothers and their families herd the deer and watch these remote territories all-year-round, whereas the chairwoman's job in Yakutsk is crucial to accessing the authorities, company headquarters and Indigenous associations to carry out necessary paperwork and networking. The combination of rural and urban members in the organizational structure and its strong ties with authorities and Indigenous associations ensure the *obshchina*'s access to various sites of negotiations, resources and flows (material and nonmaterial) regionally, nationally and internationally.

The networks in the Nordic countries are to a considerable extent meta-governed by national governments and legitimated by the participation and control by regional and local politicians (Fotel and Hanssen, 2009). This dilemma is discussed by Kaja Nan Gjelde-Bennett with reference to the mining case of Kallak in northern Sweden. Tensions have been building for years between the company, the municipal and national governments and Sámi communities in Jokkmokk municipality. Giving the Sámi a privileged claim to a part of this territory would challenge the authority of the state and the municipality, and it can also be viewed as a threat to their authority and security. As Gjelde-Bennett maintains,

> the Swedish state's interests in economic growth and protecting its sovereignty present formidable obstacles to ruling in favor of the Sámi communities. The legitimization of Indigenous rights domestically could alter the existing power structures that give preferential treatment to states operating within a neoliberal paradigm.

The state in the governance triangle

Will the states withdraw from governance, as has been frequently claimed in recent years, and will other forms of governing take their place? The findings of different cases in this book suggest the opposite. The state is leveling the playing field for Indigenous peoples in each form of governance and their concrete mixes. The state is a central actor in the governance triangle for natural resources: it is at the same time protector, promoter and regulator of natural resources, Indigenous rights and the distribution of welfare. In the everyday interactions between states, Indigenous peoples and industries, Indigenous self-determination remains a widely accepted norm, which is contested in practice. While some of the cases in this book have produced a positive outcome for Indigenous peoples—such as a withdrawn extraction license, successful agreement-making and inclusion in local debates—most of the cases represent governance failures.

From the perspective of Indigenous peoples, these cases show that governance failures result from different degrees of state compliance with international law and protection of Indigenous rights and self-determination. There is little government support for Indigenous peoples to tackle structural inequalities in interactions with governmental actors and companies, and the institutional opportunities are similarly limited for Indigenous peoples to voice their concerns, to participate and to influence development on their lands. Governance failures lead to poorly organized consultations with Indigenous organizations, while the lack of support for Indigenous participation is evident in the local and

Table 10.1 An overview of the findings from a meta-governance perspective

	Central actor(s)	Main values	Mode of governing	Conflict	Interactions	Indigenous agency	State role	Governance failures due to	Cases, including both positive and negative outcomes for Indigenous peoples
Hierarchical governance	UN mechanisms, national parliaments, governments and ministries, directorates and local administration, courts	Legitimacy, authority	Regulation, coordination	Unclear legal setting for consultations with Indigenous peoples, gap between legislation and policy implementation	Inconsistent application at national level of international law and obligations	Indigenous peoples as rights-holders in legislation and courts	The state's role as promoter of resource development vs. role as protector of Indigenous lands and rights	Lack of implementation of legal obligations, limitations of courts and national legislation to ensure protection of Indigenous rights	Kalvvatnan wind power case, Norway Pauträsk and Norrbäck wind power case, Sweden Matawa First Nations mine case, Ontario, Canada
Market governance	States and industries	Profit, time	Competition, effectiveness	Private, confidential contracts, lack of public information, closed, private interactions	Company-led interactions: IBAs, FPIC, also CSR, SLO	Rational and economic Indigenous agency; Indigenous stakeholders	State engagement in industry vs. state withdrawal from ownership, while promoting economic development by market	Unfair agreement-making processes, limited form of agency for Indigenous peoples	Tlicho mining case, Canada Adani Carmichael mine, Queensland, Australia Stekenjokk wind power plant, Sweden

(*Continued*)

Table 10.1 (Continued)

	Central actor(s)	Main values	Mode of governing	Conflict	Interactions	Indigenous agency	State role	Governance failures due to	Cases, including both positive and negative outcomes for Indigenous peoples
Network governance	Governmental and non-governmental actors, state-led networking	Trust, consensus	Argumentation	State sovereignty vs. self-determination, local and regional politics	Local to national interactions, institutional and legal opportunities for voicing Indigenous concerns	Participation, influence, representation, deliberation	Discursive and structural constraints to Indigenous agency	Diverse opportunities for normative contestation by Indigenous peoples, lack of capacity and institutional opportunities, lack of common knowledge systems	Nezhda mine case, Sakha, Russia Vedbotn aquaculture case, Norway Aquaculture in Marlborough Sounds, New Zealand Kallak mine case, Sweden

national authorities' complacency about international law and human rights principles. Scientific advice and traditional knowledge are ignored, and inadequate intersectoral coordination and multilevel policy mismatches further add to governance failure. In the following table the case results are presented from the perspective of meta-governance.

Advocates of meta-governance propose that the answer lies in reaching a normative consensus and coordination. Normative clarity in meta-governance is emphasized by professor in public organizations and management Jan Kooiman and sociologist Svein Jentoft (2009), who claim that difficult choices between values, norms and principles are easier when substantive issues are formulated and the choices inherent in them are made clear. This also requires that the process be guided by an explicit set of meta-governance principles which are deliberated by and made explicit to all concerned, public and private, in an interactive governance context (Kooiman and Jentoft, 2009). The perspective of meta-governance is one of "meta-governors" or, in other words, of political leaders and decision-makers at different levels. Political scientists Jonna Gjaltema, Robbert Biesbroek and Katrien Termeer (2020, p. 177) crystallize meta-governance as "a practice by (mainly) public authorities that entails the coordination of one or more governance modes by using different instruments, methods, and strategies to overcome governance failures." The proposed solution for meta-governance is a combination of deliberate cultivation of a flexible repertoire of responses to governance failures, a reflexive orientation about what would be acceptable policy outcomes, and regular reassessment of whether actions are producing desired outcomes. Governors should recognize that failure is likely but still continue as if success is possible (Jessop, 2011).

From the Indigenous peoples' perspectives, the situation is considerably different. As argued by the political theorist Nicolas Pirsoul (2019, p. 256), the global human rights mechanisms such as the International Labour Organization's Indigenous and Tribal Peoples Convention (ILO 169) and the United Nations Declaration on the Rights of Indigenous Peoples (UNDRIP) favor the creation of deliberative spaces between key state (and sometimes non-state) actors and Indigenous communities that are consistent with theories of deliberative democracy. However, the consultations that do take place suffer from a deliberative deficit. Pirsoul uses New Zealand and Colombia as case studies, and argues that consultation would better help respect the rights of Indigenous peoples if they were consistent with the political ideals that inform deliberative democratic theory. This deliberative deficit is obvious in many of our cases, too.

Due to the governance failures stemming from the deliberative deficit, the Indigenous peoples' future seems to entail a continuous struggle to advance their rights and interests in natural resource governance. They will use hybrid strategies to promote Indigenous self-determination in national legal processes, company-led agreement-making and local networking. By providing alternative future imaginaries in contrast to often neoliberal, extractive economic development, Indigenous peoples will continue to contest the norms. To turn governance failures into successes requires most of all institutional sites for deliberation and normative contestation by Indigenous peoples in their interactions with states and extractive industries. This will take place one struggle at a time, and the states will still have a central role in leveling the playing field for Indigenous peoples in the governance triangle of natural resources in the Arctic and beyond.

References

Alcantara, C. and Morden, M. (2019) "Indigenous multilevel governance and power relations," *Territory, Politics, Governance*, 7(2), pp. 250–264.

Berg-Nordlie, M. (2018) "Substitution in Sápmi: meta-governance and conflicts over representation in regional Indigenous governance" in S. Kropp, A. Aasland, M. Berg-Nordlie, J. Holm-Hanssen and J. Schuhmann (eds.) *Governance in Russian regions: a policy comparison*. Cham: Palgrave MacMillan, pp. 189–218.

Fotel, T. and Hanssen, G.S. (2009) "Meta-governance of regional governance networks in Nordic countries," *Local Government Studies*, 35(5), pp. 557–576.

Gjaltema, J., Biesbroek, R. and Termeer, K. (2020) "From government to governance … to meta-governance: a systematic literature review," *Public Management Review*, 22(12), pp. 1760–1780.

Humpage, L. (2010) "Revisioning comparative welfare state studies: an 'Indigenous dimension'," *Policy Studies*, 31(5), pp. 539–557.

Jessop, B. (1998) "The rise of governance and the risks of failure: the case of economic development," *International Social Science Journal*, 50, pp. 29–45.

Jessop, B. (2011) "Meta-governance" in M. Bevir (ed.) *The SAGE handbook of governance*. London: SAGE, pp. 106–123.

Junka-Aikio, L. and Cortes-Severino, C. (2017) "Cultural studies of extraction," *Cultural Studies*, 31(2–3), pp. 175–184.

Kooiman, J., Bavinck, M., Jentoft, S. and Pullin, R. (eds.) (2005) *Fish for life: interactive governance for fisheries*. Amsterdam: Amsterdam University Press.

Kooiman, J. and Jentoft, S. (2009) "Meta-governance: values, norms and principles, and the making of hard choices," *Public Administration*, 87, pp. 818–836.

Kropp, S. and Schuhmann, J. (2016) "Governance networks and vertical power in Russia: environmental impact assessments and collaboration between state and non-state actors," *East European Politics*, 32(2), pp. 170–191.

Kuokkanen, R. (2019) "At the intersection of Arctic Indigenous governance and extractive industries: a survey of three cases," *The Extractive Industries and Society*, 6(1), pp. 15–21.

La Cour, A. and Aakerstroem Andersen, N. (2016) "Meta-governance as strategic supervision," *Public Performance & Management Review*, 39(4), pp. 905–925.

Larsen, R.K. and Raitio, K. (2019) "Implementing the state duty to consult in land and resource decisions: perspectives from Sami communities and Swedish state officials," *Arctic Review on Law and Politics*, 10, pp. 4–23.

Lawrence, R. and Moritz, S. (2019) "Mining industry perspectives on Indigenous rights: corporate complacency and political uncertainty," *The Extractive Industries and Society*, 6(1), pp. 41–49.

MacDonald, F. (2011) "Indigenous peoples and neoliberal 'privatization' in Canada: opportunities, cautions and constraints," *Canadian Journal of Political Science/Revue canadienne de science politique*, 44(2), pp. 257–273.

Meuleman, L. (2008) *Public management and the meta-governance of hierarchies, networks and markets: the feasibility of designing and managing governance style combinations*. Heidelberg: Springer.

Meuleman, L. (2019) *Meta-governance for sustainability: a framework for implementing the sustainable development goals*. London and New York: Routledge.

Pirsoul, N. (2019) "The deliberative deficit of prior consultation mechanisms," *Australian Journal of Political Science*, 54(2), pp. 255–271.

Smith, D. (2008) "From collaboration to coercion: a story of governance failure, success and opportunity in Australian Indigenous affairs" in J. O'Flynn and J. Wanna (eds.) *Collaborative governance: a new era of public policy in Australia*. Canberra: ANU Press, pp. 75–92.

Sørensen E. and Torfing J. (2007) "Introduction. Governance network research: towards a second generation" in E. Sørensen and J. Torfing (eds.) *Theories of democratic network governance*. London: Palgrave Macmillan, pp. 1–21.

Strakosch, E. (2015) *Neoliberal Indigenous policy: settler colonialism and the "post-welfare" state*. London: Palgrave Macmillan.

Tysiachniouk, M.S., Petrov, A.N. and Gassiy, V. (2020) "Towards understanding benefit sharing between extractive industries and Indigenous/local communities in the Arctic," *Resources*, 9(4), pp. 48–53.

Willow, A.J. (2019). *Understanding extrACTIVISM: culture and power in natural resource disputes*. New York, NY: Routledge.

Wilson, E. and Stammler, F. (2016) "Beyond extractivism and alternative cosmologies: Arctic communities and extractive industries in uncertain times," *The Extractive Industries and Society*, 3(1), pp. 1–8.

Index

Page numbers in **bold** indicate tables, page numbers in *italic* indicate figures.

Printed in the United States
by Baker & Taylor Publisher Services